Acoustic Emission Technology for High Power Microwave Radar Tubes

*Authored by Narayan R. Joshi,
Ayax D. Ramirez, Stephen D. Russell
and David W. Brock*

Published in London, United Kingdom

IntechOpen

Supporting open minds since 2005

Acoustic Emission Technology for High Power Microwave Radar Tubes
http://dx.doi.org/10.5772/intechopen.72298
Authored by Narayan R. Joshi, Ayax D. Ramirez, Stephen D. Russell and David W. Brock

Contributors
Ayax D. Ramirez, Narayan R. Joshi, Stephen D. Russell, David W. Brock

First published in London, United Kingdom, 2019 by IntechOpen
IntechOpen is the global imprint of INTECHOPEN LIMITED, registered in England and Wales, registration number: 11086078, The Shard, 25th floor, 32 London Bridge Street
London, SE19SG – United Kingdom
Printed in Croatia

British Library Cataloguing-in-Publication Data
A catalogue record for this book is available from the British Library

Additional hard copies can be obtained from orders@intechopen.com

Acoustic Emission Technology for High Power Microwave Radar Tubes
Authored by Narayan R. Joshi, Ayax D. Ramirez, Stephen D. Russell and David W. Brock
p. cm.
Print ISBN 978-1-78985-455-8
Online ISBN 978-1-78985-456-5

For EU product safety concerns:
IN TECH d.o.o., Prolaz Marije Krucifikse Kozulić 3, 51000 Rijeka, Croatia, info@intechopen.com or visit our website at intechopen.com.

Meet the authors

Narayan R. Joshi received his MSc in Electronics from Poona University, India. He received his MS and PhD in Mechanics and Materials Science from Johns Hopkins University, MD. He has worked as the Head of the Dual Engineering Degree at the University of Maryland, Baltimore, Lamar University, TX, and Prairie View A&M University, TX. His academic teachings include elastic wave propagation, materials science, metallurgy, mechanical vibrations, and advanced non-destructive evaluation methods. His experimental research includes ultrasonics, acoustic emission, X-ray diffraction, and optical materials (lasers without inversion). Dr. Joshi has worked for General Electric, Oak Ridge National Laboratories, Southwest Research Institute, and Texas Instruments. He has about 60 research publications to his credit. He is ASNT Level III in UT, PT, MT, and ET. He is a fellow of the American Society for Non-Destructive Testing. He holds three patents.

Mr. Ayax D. Ramirez is a senior physicist and manager of the Basic and Applied Research Division at SPAWAR Systems Center Pacific (SSC Pacific). He manages and leads the technical direction of over 100 scientists and engineers, with the mission to provide scientific and technological solutions in the areas of cryogenic exploitation of RF signals, quantum information technologies, cyber security, non-linear dynamics, environmental sciences, alternative energy sources, electromagnetic sensors, human/machine interactions, and other technologies relevant to the US Navy. Mr. Ramirez's career with the US Navy spans over 20 years and his work has been recognized with numerous awards, including the Lauritsen-Bennett Award for Excellence in Science, the Navy Meritorious Civilian Service Award, the Federal Lab Consortium Excellence in Technology Transition Award, the Hispanic Engineer National Achievement Luminary Award, the Meritorious Service Award from the National Organization for Mexican American Rights, the Black Engineer of the Year Award, Science Technology, Engineering and Mathematics (STEM) Trailblazer Award, the SSC Pacific Science and Technology Award, and the SSC Pacific Leadership Award. Mr. Ramirez holds over 30 US patents and has authored or coauthored over 40 publications in peer-reviewed journals, conference proceedings, and technical reports. Mr. Ramirez received his MS in Physics from San Diego State University. He is a member of the National Academy of Inventors and the American Physical Society. In addition, Mr. Ramirez is an Adjunct Professor of Physics at Southwestern College in Chula Vista, CA, where he has been teaching for over 20 years.

Dr. Stephen Russell, Senior Executive Service (SES), is Director of Science and Technology and Chief Technology Officer at Space and Naval Warfare Systems Command (SPAWAR), and Head of the SPAWAR Systems Center Pacific (SSC Pacific) Cyber/S&T Department where he leads a highly technical team of 1000+ personnel and influences $1.2B supporting R&D in the command and control, communications, computers, intelligence, surveillance and reconnaissance, cyber, and space domains. Over his 30+ year career supporting the Department of the Navy he has been the recipient of many awards, including AFCEA's International Gold Medal in Engineering, SSC Pacific's Lauritsen-Bennett Award for Excellence in Science, the Navy Meritorious Civilian Service Award, and the Federal Lab Consortium Excellence in Technology Transition Award. He holds over 120 US and foreign patents and over 20% of his IP portfolio has been commercially licensed. He has authored or coauthored over 70 articles in peer-reviewed journals, conference proceedings, and technical reports and serves on the Editorial Board of the Naval Science and Technology FUTURE FORCE Magazine. Dr. Russell received his PhD, MS from the University of Michigan and his BS from SUNY Stony Brook. He is a Fellow of the National Academy of Inventors, Senior Member of SPIE, and member of the American Physical Society.

David W. Brock, scientist in the Intelligence, Surveillance and Reconnaissance Electromagnetic Systems Technology Branch at SPAWAR Systems Center Pacific, received his PhD in Physics from the University of Texas at Austin in 1981, and his MS in Physics from the University of Colorado at Boulder in 1975. He received BS degrees in Physics and Mathematics at the New Mexico Institute of Mining and Technology in 1972. Prior to joining SPAWAR Systems Center, Dr. Brock worked as a research assistant at the University of Texas (1977–1981) and the University of Colorado (1975). He has worked at SPAWAR Systems Center since 1981 initially in the High Power Microwave Tube Branch and later in the Microwave/Millimeter Wave Technology Branch. He has carried out research work in areas including high-power microwave tubes, the design and building of radar systems, microwave receivers, and antennas. Dr. Brock has published several technical articles, and has 17 patents issued or pending. He is a member of the National Academy of Inventors.

Contents

Preface

Acoustic emission (AE) refers to the generation of transient elastic waves by a sudden redistribution of stress in a material. When a structure such as a pressure vessel is subjected to an external stimulus, for instance a change in pressure, load, or temperature, localized sources trigger the release of energy in the form of stress waves that propagate to the surface. Acoustic emission sensors, coupled to the surface of the structure under test, record the emitted stress waves. The stress applied to materials under test could be tensile, compressive, or shear. The transient elastic stress waves of AE have frequencies ranging from 20 kHz to 1 MHz.

Experimental research work shows that AE activity could be generated by many different mechanisms, such as mechanical deformation, fracture, crack propagation, dislocation motion and multiplication, twin formation, phase transformation, corrosion, friction, and internal magnetic processes. Mechanical loading is not the only way to generate AE activity (phonon signals). Thermal shock loading and electrical sparking have also caused AE activity. Generation of AE activity during chemical reactions has also been observed. It was realized quite early in the research that AE activity appears in two types: burst signals and continuous signals.

In recent years, it was realized that high-power microwave radar tubes needed simple non-destructive testing (NDT) techniques for in-situ performance monitoring. Microwave vacuum tubes used in high-power radar and communications systems have a lifetime of a few thousand active hours before refurbishment is required. When a costly tube fails, it is generally impossible to determine the sequence of events that contributed to its failure. Each year, nearly $100 million is spent replacing high-power microwave tubes in the fleet. In many cases, estimated at over 25%, tubes that operate perfectly are inadvertently replaced because of insufficient in-situ monitoring equipment available to diagnose specific problems within the system. This results in high maintenance and refurbishing costs.

Currently, microprocessor-based systems with as many as eleven sensors are designed to monitor tube performance, provide tube protection, and record a comprehensive tube failure history. A major limitation with these systems is the small amount of time available during the tube interpulse period for data buffering and fault analysis. These monitoring systems work well if the microwave tube is operated at rates of 200 pulses per second or less. Typically, these tubes are operated at up to 1000 pulses per second, with a pulse duration of 1 microsecond. Increasing the data acquisition speed will, in this case, make the situation worse, since it will increase the amount of data that must be transferred and analyzed during the small time interval available. Therefore, although microprocessor-based monitoring systems have merits, they also have limitations. So, there is a need for a different non-destructive examination (NDE) solution that would overcome the limitations created by high-pulse repetition rates.

Research over the past several years has established the effectiveness of acoustic emissionAE-based sensing methodologies for machine condition analysis and process monitoring. The success of AE NDT for many different applications directed us to try AE as an advanced NDT method for in-situ performance monitoring of high-power microwave tubes. These experiments proved the feasibility of AE as a real-time nondestructive testing technique, which is much simpler than the multisensor microprocessor-based systems. This book presents the results of this new application of AE testing to different types of microwave high-power radar tubes.

Narayan R. Joshi, PhD

Acoustic Emission Technology for High-Power Microwave Radar Tubes

Narayan R. Joshi, Ayax D. Ramirez, Stephen D. Russell and David W. Brock

Abstract

Microwave tubes used in high-power radar and communications systems are expensive and have an operating life of a few thousand hours. When one fails, it is generally impossible to determine the sequence of events that contributed to its failure. Previous investigators have designed microprocessor-based systems with as many as 11 sensors to monitor tube performance, provide tube protection, and record a comprehensive tube failure history. These systems are limited by the small amount of time available during the tube's interpulse period for data buffering and fault analysis. They work well if the microwave tube is operated with 200 or fewer pulses per second. However, many tubes are operated at up to 1000 pulses per second. In this effort, an alternative nondestructive testing technique using acoustic emission (AE) was used for in-situ monitoring of normal and abnormal performance of radar tubes, including a magnetron, a klystron, and a traveling wave tube amplifier. This technique captures changes in radio frequency (RF) output pulses due to irregular operation and it is a real-time instantaneous in-situ indicator of the performance of microwave radar tubes. It also offers the possibility of developing built-in prognostic capabilities within the radar system to provide advanced warning of a system malfunction. Understanding the sequence of events leading to a tube failure allows for better maintenance, extends the operating life of the system, and results in significant cost avoidance.

Keywords: acoustic, emission, technology, microwave, radar

Chapter 1

Acoustic Emission

1.1. Acoustic emission phenomena

For centuries, woodsmen have listened to the groaning sounds emitted by trees as they fall and taken it as a warning to stand clear. Alchemists during the Middle Ages recognized acoustic emissions given by tin, zinc, and iron during metal working. Acoustic emission (phonon signals) and microseismic activity are naturally occurring phenomena. Almost all materials emit sound or acoustic emission (AE) when they are sufficiently stressed. Wood and rock produce AE signals in the audible frequency range when stressed. It was conjectured that there existed the possibility of AE signal generation in the ultrasonic frequency range during deformation of materials. In 1950, Joseph Kaiser reported the first comprehensive investigation into acoustic emission phenomenon in conventional engineering materials using electronic equipment and a tensile testing machine. Kaiser also observed that AE activity was irreversible. Acoustic emissions are not generated during the reloading of a material until the stress level on it exceeds its previous high value. This AE irreversible phenomenon is now known as the Kaiser effect [1]. He also proposed a distinction between burst and continuous AE activity.

In the middle of the 1950s in the United States, Schofield and Tatro investigated AE phenomena. They found that the emissions from metals were primarily due to dislocation motions accompanying plastic deformation, rather than being entirely due to grain boundary sliding in polycrystalline materials as proposed by Kaiser. The use of piezoelectric transducers and electronic amplifiers to observe high frequency AE activity gradually led to the definition of acoustic emission. Acoustic emission thus refers to the generation of transient elastic stress (strain) waves due to the rapid release of energy from a localized source within a material undergoing some kind of deformation. The stresses applied to materials under test could be tensile, compressive, or shear. AE activity on a large scale appears during plastic deformation of materials. The transient elastic stress waves of AE have frequencies ranging from 20 kHz to 1 MHz. Experimental research work showed that AE activity could be generated by many different mechanisms such as mechanical deformation, fracture, crack propagation, dislocation motion and multiplication, twin formation, phase transformation, corrosion, friction, and internal magnetic processes. Mechanical loading is not the only way to generate AE activity. Thermal shock loading and electrical sparking can also cause AE activity. The generation of AE activity during chemical reactions has also been observed. It was realized quite early that AE activity appears in two types, burst and continuous signals.

1.2. AE signal characteristics

If an AE signal, as seen on an oscilloscope, consists of distinct pulses that are distinguishable from the background noise, the signal can be characterized as a burst emission. If a signal is not resolvable into individual bursts, then it is called a continuous emission [2]. **Figure 1.1** shows oscilloscope displays of these two types of AE signals. The burst type of the signal in the amplitude versus time format can be described by many parameters, such as rise time, rise time slope, peak amplitude,

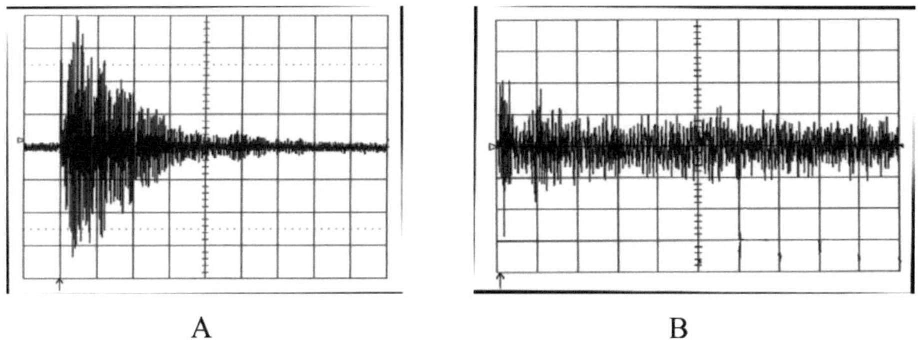

Figure 1.1.
(A) Burst acoustic emission (AE) and (B) continuous AE.

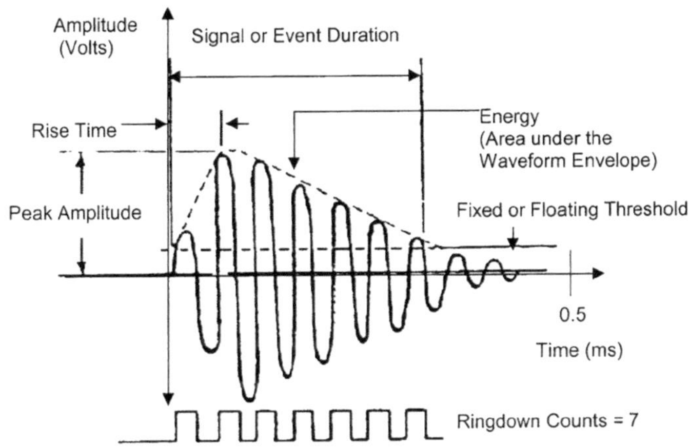

Figure 1.2.
Different parameters of acoustic emission signals.

ring-down count, energy envelope, and event (or signal) duration. The typical burst type waveform and its parameters are shown schematically in **Figure 1.2**. Continuous emission is a qualitative term applied to acoustic emission when the burst or pulses are not discernible. Background noise is the signal in the absence of any acoustic emission events. It has electrical and mechanical origins.

1.3. Instrumentation to examine AE signals

All AE signals are mechanical signals. It is necessary to convert them into electrical signals to display, record, or analyze them in real time. This function is performed by an acoustic emission sensor coupled to a specimen undergoing stress testing using a suitable couplant made of viscous liquid. A simple system to study acoustic emissions generated by a specimen subjected to changing stress is shown schematically in **Figure 1.3**. The critical component in the instrumentation system is the sensor chosen to pick up the AE signal.

These sensors are usually piezoelectric transducers that can be either narrow band or broad band depending upon the investigation to be performed. Capacitive sensors are used in laboratories for their special characteristics of flat frequency response. If no contact with the specimen under test is desired, then an electromagnetic sensor is the choice. However, its sensitivity is lower than that of a piezoelectric sensor. AE signals can also be detected by optical methods as described

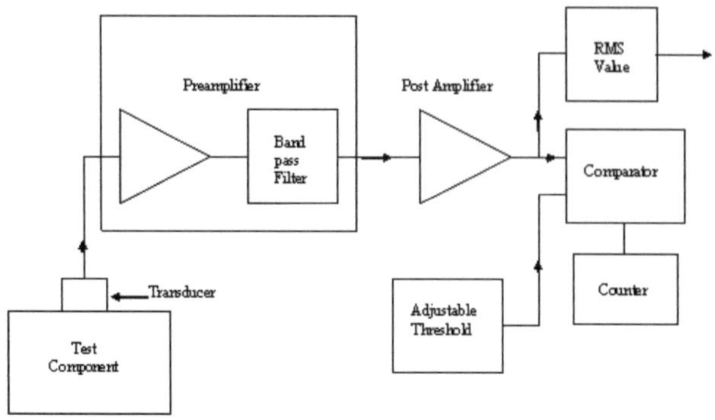

Figure 1.3.
Schematic diagram of a simple acoustic emission system.

by Palmer and Green [3]. The sensitivity of a sensor depends upon the frequency, wave type, and sensor design. It can be specified in terms of an output voltage produced by a given input quantity. In the case of simple AE signal detection by a piezoelectric transducer system, for example, a sensor + any signal conditioning amplifier + cables + some readout device, the sensitivity is expressed as the ratio of (electrical output signal)/(mechanical input).

Acoustic emission sensor calibration is necessary in order to obtain a plot of frequency versus sensitivity. In frequency calibration, the arrangement should reproduce an AE type situation so that it is valid for all AE measurements performed. Because AE may be continuous or discrete, some attempts have been made to generate standard pulses which for discrete ones correspond to the delta function, whereas in the case of the continuous emission, to the white noise. As a discrete pulse source, the breaking of glass capillary proposed by Hsu and Breckenridge [4] works well. For calibration of equipment designed for registering continuous AE, McBride and Hutchinson [5] suggested a method based on the outflow of gas flux from a nozzle. A modified version of artificial AE pulse source called a Hsu-Nielsen source appears to meet the requirements of a repeatable AE source, characterized with simplicity of construction and ease of application under industrial conditions [6].

Referring to **Figure 1.3**, a weak signal, generated by the sensor, of approximately 1 μV amplitude, is fed to a preamplifier located close to the sensor in order to minimize attenuation of the signal on the cable before amplification. A band-pass filter is used to filter out unwanted noise. The electrical AE signal is further amplified by a postamplifier located away (in certain experimental situations) from the sensor coupled to an object under testing. The output of the postamplifier feeds into averaging circuitry to determine its moving time average RMS value and into a comparator. A threshold voltage is sent into the comparator from an adjustable source. This threshold voltage can either be fixed or be set to automatically float above the peak value of the background noise. This threshold voltage is shown in **Figure 1.2**. Since the threshold is set to some level above the background noise, the comparator will put out pulses and register the signal only when the postamplifier output exceeds the background noise. The output of the comparator is fed into a counter.

1.4. Methods of AE data analysis

1.4.1 Ring-down count and event count

At this stage, the AE electrical signals can be counted, integrated, captured, or characterized in many different ways. The simple method of ring-down counts measures the number of times the amplitude of a damped sinusoidal signal exceeds a chosen threshold voltage within a single event. The total cumulative counts and the count rate are also used to characterize AE activity. For the purpose of comparison of ring-down counts in different AE events, gain values on amplifiers must be held at constant settings chosen suitably for the experiment. The number of AE events and the rate of occurrence of these events during a test have been correlated, as summarized by Bassim and Houssny-Emam [7], to various physical variables like load, strain, and the number of cycles in fatigue testing and to the stress intensity factor in fracture mechanics. The counting of acoustic events is simple and convenient. Although audible signals can be counted by ear, signals of ultrasonic frequencies are counted by using high-speed comparator integrated circuits and data loggers. As discussed previously, a threshold reference voltage is set to some level above the background and fed into the comparator, so that when an emission occurs that exceeds this level, the signal is registered. Circuits include a delay timer to ensure that only a single count is registered per incoming oscillation signal. Counting AE events may be inadequate for some quantitative applications because AE event counting is influenced by many factors like resonances in the test structure, sensor resonance and damping, nature of coupling between the sensor and the test structure, and the rate of loading the structure.

1.4.2 Energy and RMS measurements

AE signals differ in both peak amplitudes and total energies. Hence, there is a second approach for characterizing AE signals, related to the measurement of their energy content.

Acoustic emission is attributed to the release of energy from any structure under stress. So the measured energy content of an AE signal is an indication of this energy release. Instruments are commercially available that approximate the total energy in each emission. Another useful approach for fast emitting systems is to integrate the output, using either a peak level detection device or a true RMS meter. Interference from intense but very short-lived transients resulting from electrical switching can be reduced by considering the total energy of acoustic events, rather than their maximum amplitudes alone. **Figure 1.4** shows the steps in energy measurement. The energy measurement is a procedure for measuring the area under the envelope of the square of the signal using digital integration. Beattie [8] demonstrated the advantage of using the energy analysis method over that of ring-down counts, especially in measuring the high energy events where ring-down counts failed to register the proper relative magnitudes. Kline [9] has shown, for a test of constant strain rate deformation of lead-tin alloy, that in the beginning of the test, the count activity was stronger than RMS response due to burst type AE activity, while in the later part of the test, RMS response peaked due to continuous emission. However, according to Beattie, the energy and power measured are those of the electrical signal generated by the transducer and neither is proportional to the energy of the acoustic wave or the AE source.

1.4.3 Analysis of AE signal amplitudes

Amplitude distribution analysis is another way of characterizing acoustic emission from a test object. Since intensity is proportional to the square of the amplitude, the amplitude of an AE signal is indicative of the energy involved in generating it. In other words, it is indicative of the intensity of the damage in the material that caused the signal to occur. The amplitudes of detected signals are dependent on the position of the transducer relative to the emitting source or sources and the coupling between the transducer and the test object. AE signal

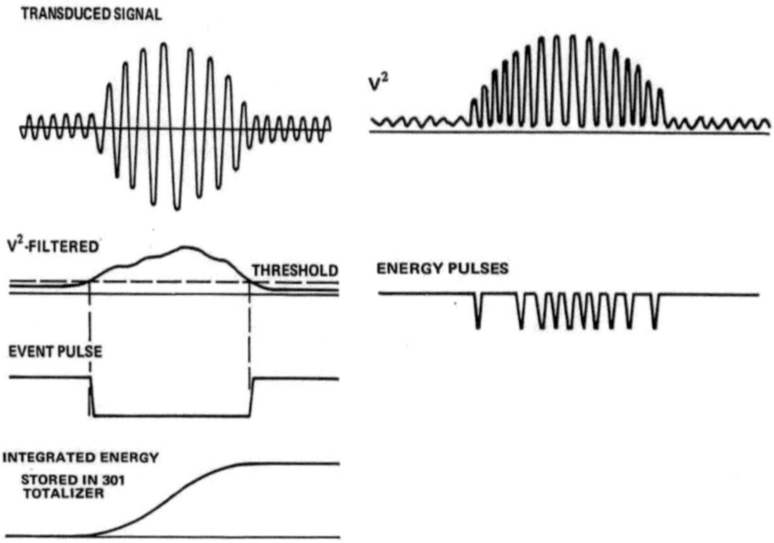

Figure 1.4.
Steps in energy measurement.

amplitudes are attenuated and greatly modified by wave propagation mechanisms which are functions of properties of the propagating medium and its geometry. The peak signal amplitude is the maximum absolute amplitude of an AE burst. It could be positive or negative. Disparity in magnitudes between positive and negative peak amplitudes is typical of AE signals [10]. However, in a controlled test settings where factors such as coupling of transducer, position of transducer, wave propagation medium, and gains on both amplifiers are maintained constant, amplitudes generated by two distinct mechanisms triggered from outside in the test object could be compared to obtain reliable information about them.

1.4.4 Acoustic emission spectral analysis

With advances in computer and related technologies, AE signal spectral analysis can now be performed by applying fast Fourier transform (FFT) on each time domain signal. The transformed data are usually viewed as a power spectrum. Power observed within various frequency ranges can be summed to provide better quantitative information. Success of the spectral analysis of the AE signal varies and appears to be limited. This is because the detected AE event is the composite result of the source event modified by the input/output instrumentation system and can be described conveniently with different transfer functions such as the specimen transfer function, the transducer transfer function, and the signal processing transfer function [11]. Even when a capacitive transducer or optical probe is used to faithfully detect the surface displacement waveform characteristic of an acoustic emission source, the waveform and hence its frequency spectrum appear to undergo severe modification as it reflects back and forth in an actual specimen [12].

1.5. New techniques to study acoustic emission phenomenon

It has been observed over a number of years that recorded AE signals show the existence of different waveform categories. Digital signal analysis techniques have made the procedure for classifying different waveforms and correlating them according to underlying microfracture processes less laborious. Ono and Huang [13] provided the basic concept and applied it to the characterization of composite microfractures. They used the ICEPAK software package which includes five statistical and neural network classifiers. With its incorporation of artificial intelligence in the form of neural networks, statistical pattern classifiers, and fuzzy logic, super ICEPAK is a sophisticated program for multiparameter automatic data interpretation. Ono [14] emphasized that the effects of wave dispersion and attenuation need to be considered carefully in employing these procedures when the distances between AE source and sensor vary widely. Wideband sensors and electronics are also needed to discriminate different signal features since narrow-band sensors drastically alter waveform and spectral features. The power spectrum allows descriptors to be extracted from it, such as median frequency and bandwidth. Betteridge et al. [15] used pattern recognition techniques to show the similarity between signals from various subsets of 34 different acoustically emissive chemical reactions. Many pattern recognition techniques that are now commonplace for analyzing and classifying more conventional acoustic emission data are also applicable to chemical (and other) sources of AE. Recently, Sachse and his coworkers [16] applied an adaptive learning system to the characterization of acoustic emission phenomenon.

A powerful new tool of wavelet transformation is also making inroads into the analysis of AE signals. Since AE signals are of the burst type, classical analysis of them by Fourier transformation is not the best approach. Wavelet transform allows the determination of the frequency spectrum as a function of time and frequency using short waveform segments or wavelets as the basis functions. With a good selection of a mother wavelet function, the basis function will be more identical to the burst form of the signal than the sine base function of Fourier transformation is the resulting mapping of wavelet coefficients in the frequency-time coordinate plane provides a more informative characterization of transient signals than the power density spectra from the usual Fourier transform. Suzuki et al. [17] performed wavelet transform on AE signals from the longitudinal glass fiber-reinforced composite material sample under tensile loading. The spectrograms obtained were classified into four types and compared to known fracture dynamics results with excellent correlation.

1.6. Emission source location, orientation, and inverse problem

Determining flaw location in a structure in real time is one of the most important applications of AE as a nondestructive testing technique (NDT). Flaw location is based on the measurement of relative arrival times of an AE signal at several sensors attached to the structure under load test and is therefore readily applicable to cases where burst emissions are generated. The linear location mode of investigation is used for a linear structure, the triangular mode for a shell type surface structure, and the spatial mode for source location in three dimensions. The actual location of an AE source in a structure can be verified if a signal from a computer simulation appears on the display at the same place as a suspected source of acoustic emission. Calibration signals are used to adjust the gain of all sensors to the same level, determine the attenuation in the test structure, and provide a relative basis for comparison on AE signals. Once a source has been located by AE, the next task would be to determine its size and orientation.

There are different ways to deal with this part of the problem, sometimes called the inverse problem. Standard nondestructive testing techniques like ultrasonics or X-rays could be used to locate body defects and magnetic particle or liquid penetrant to delineate surface flaws. For some well-defined systems, a deconvolution technique could be used, but in some cases, the crack orientations can cause the deconvolution method to produce erroneous results. In recent years, further progress has been made in determining the orientation of cracks. A new method of moment tensor scheme uses five transducers mounted in different locations on a structure to record an AE event. Next a "moment tensor" is derived from the digital records of this signal. Eigenvector analysis enables the tensor to be separated into a mean component, a shear component, and a component known as the compensated linear vector dipole. It is possible to classify an AE source as either a tensile crack or a shear crack from the ratios of these three components [18].

References

[1] Liptai RG, Harris DO, Tatro CA. An introduction to acoustic emission. In: Acoustic Emission, STP 505, ASTM. West Conshohocken, PA, USA: American Socity of Testing Materials. 1971

[2] Bray DE, McBride D. Nondestructive testing techniques. In: Acoustic Emission Technology. New York: John Wiley and Sons. 1992. pp. 345-347. Chapter 12

[3] Palmer CH, Green RE. Materials evaluation by optical detection of acoustic emission signals. Materials Evaluation. 1977;**35**(10):107-112

[4] Hsu NN, Breckenridge FR. Characterization and calibration of acoustic emission sensors. Materials Evaluation. 1981;**39**:60-69

[5] McBride SL, Hutchinson T. Absolute calibration of the helium gas jet noise source. Canadian Journal of Physics. 1978;**504**:54

[6] Jerzy S, Marcin L. Measurement and analysis of acoustic emission standard pulses generated in Hsu-Nielsen source. Archives of Electrical Engineering. 1998;**XLVII**(1):13-24

[7] Bassim MN, Houssny-Emam M. Time and frequency analysis of acoustic emission signals. In: Matthews JR, editor. Acoustic Emission. New York: Gordon and Breach Science Publishers; 1983. pp. 139-163

[8] Beattie AG. Energy analysis in acoustic emission. Materials Evaluation. 1976;**34**:73-78

[9] Kline RA. Acoustic emission signal characterization. In: Matthews JR, editor. Acoustic Emission. New York: Gordon and Breach Science Publishers; 1983. pp. 105-137

[10] Beattie AG. Acoustic emission, principles and instrumentation. Journal of Acoustic Emission. 1983;**2**(1/2):95-128

[11] Simmons JA, Clough RB. Theoretical aspects of acoustic emission spectral measurements. In: Proceedings of the Eighth World Conference on Nondestructive Testing. Cannes, France; 1976

[12] Green RE Jr. Basic wave analysis of acoustic emission. In: Steinchcomb WW, editor. Mechanics of Nondestructive Testing. New York, NY: Plenum Publishing Corporation; 1980. pp. 55-76

[13] Ono K, Huang Q. In: Kishi T et al., editors. Progress in Acoustic Emission VII. Tokyo: Japan Society NDI; 1994. pp. 69-78

[14] Ono K. Recent development in acoustic emission. Journal of Acoustic Emission. 1997;**15**(1-4):S95-S102

[15] Betteridge D, Joslin MT, Lilley T. Acoustic Emissions from Chemical Reactions. Analytical Chemistry. 1981;**53**:1064-1073

[16] Grabec I, Sachse W, Govekar E. Solving AE problems by neural networks. In: Sachse W et al., editors. ASTM STP 107, Acoustic Emission: Current Practice and Future Directions. West Conshohocken, PA, USA: American Socity of Testing Materials. 1991. pp. 165-182

[17] Suzuki H, Kinjo T, Hayashi Y, Takemoto M, Ono K. Wavelet Transform of Acoustic Emission Signals. Journal of Acoustic Emission. 1996;**14**:69-84

[18] Ohtsu M. Source kinematics of acoustic emission based on a moment tensor. NDT International. 1989;**22**(1):14-20

Chapter 2

Practical Applications of Acoustic Emission Technology

Applications of acoustic emission (AE) technology are growing rapidly, with AE instrumentation undergoing improvement through several generations of refinement, and have reached the stage where they are now ready to offer high levels of performance and user convenience. A short review is given here of some applications. Interested readers can find vast literature on this subject in the American Society for Nondestructive Testing (ASNT) Handbook on Acoustic Emission Testing [1].

2.1. Materials science and metallurgy

In experiments on copper single crystals, Pollock [2] observed that at very low strains, the material supports easy gliding, with dislocations traveling over long distances. The source events are thus of long duration, and the associated spectrum of acoustic emissions falls off between 70 and 350 kHz. At higher strains, the easy movement of dislocations is blocked and the distances traversed become shorter. These shorter duration events have spectra extending well above 350 kHz.

For a polycrystalline 7075-T6 aluminum unflawed tensile specimen, Dunegan and Harris [3] reported very nice data of the acoustic emission rate as a function of strain. The experiment showed that the acoustic emission began to reach a maximum just before the proportional limit of the stress-strain curve. An interesting superimposition on the AE data, a fit of Gilman's mobile dislocation model, gave mobile dislocation density as a function of plastic strain. Although the agreement for aluminum specimen was excellent for this superimposition, experiments on Fe-3%Si specimens showed a different picture of relatively poor agreement.

Most AE research and development has occurred in metallurgy and materials science. This is because early AE research showed that emissions are generated in metals by dislocation motion, grain boundary sliding, crystal twinning, phase transformations, crack initiation, crack growth from fatigue or corrosion, and cracking of second-phase particles. Metal fatigue is a gradual deterioration of the crystal structure by cyclic stresses and is responsible for the abrupt breaking of metallic machine components subjected to cyclic stresses, without showing external signs of any permanent deformation to herald the fracture. Much effort has gone into the theoretical modeling of fatigue phenomena, as well as into the nondestructive testing of fatigue progress, to obtain a kind of warning mechanism prior to crack initiation into an unflawed specimen subjected to cyclic stresses. Both scientists and engineers want to use the AE technique to monitor the progress of fatigue damage. Looking now at another nondestructive testing technique using ultrasonics, the ultrasonic attenuation detection of fatigue damage was reported as early as 1972 by Joshi [4]. They successfully showed (**Figure 2.1A**), through a large number of tests, that the attenuation of ultrasonic pulses propagating through aluminum

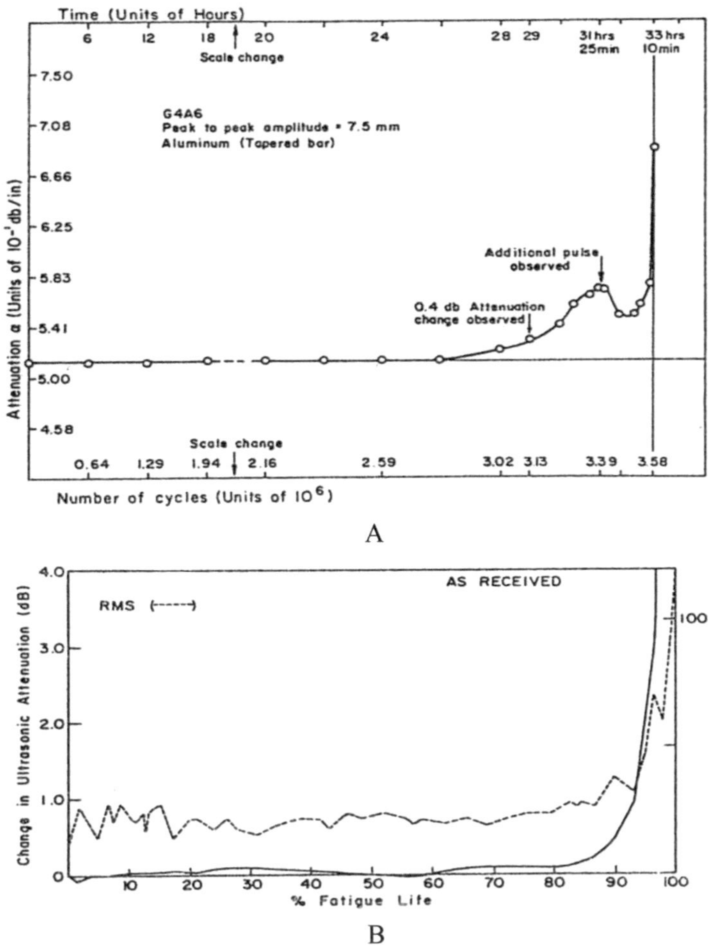

Figure 2.1.
(A) Typical plot of ultrasonic attenuation versus the number of fatigue cycles for aluminum (after Joshi and Green [5]). (B) Typical plot of change in ultrasonic attenuation and true root mean square voltage of acoustic emission versus fatigue life for 7075-T651 aluminum (after Green and Duke [6]).

or steel bars subjected to cyclic loading initially remained constant, then increased slowly with the number of fatigue cycles, and finally increased catastrophically just prior to fracture of the test specimen [5].

In these tests, ultrasonic attenuation was a sensitive indicator of fatigue damage and indicated that failure was imminent before the conventional ultrasonic pulse-echo monitoring could detect an echo due to energy reflected from a crack. Ultrasonics is an active technique, while acoustic emission is a passive technique. So, acoustic emission could be used to listen to a kind of crying signal given out by a tired specimen subjected to cyclic stresses. **Figure 2.1B** shows the results of AE testing of an aluminum specimen during a fatigue experiment. The experimental apparatus (**Figure 2.2B**) used in joint ultrasonic and acoustic emission detection of fatigue damage by Green and Duke [6] was similar to the one used by Joshi and Green (**Figure 2.2**), only for ultrasonic detection of fatigue damage. The two techniques could serve to be complementary to each other.

Harris et al. [7] discussed a new method in order to test the integrity of a structure subjected to fatigue loading. They used acoustic emission in conjunction with periodic proof stressing which provided a means of detecting the

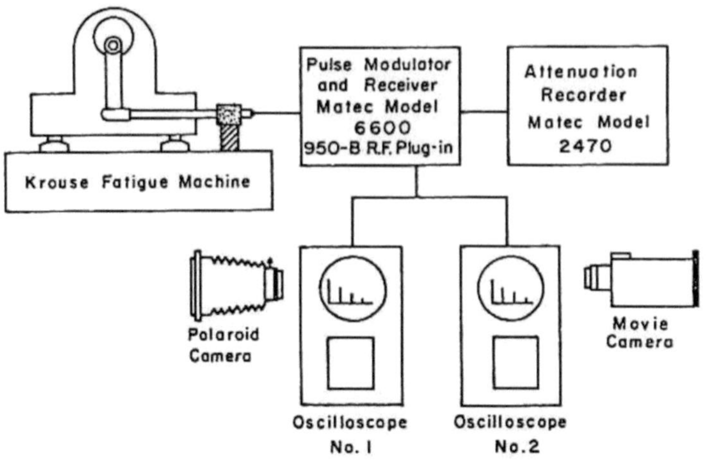

A

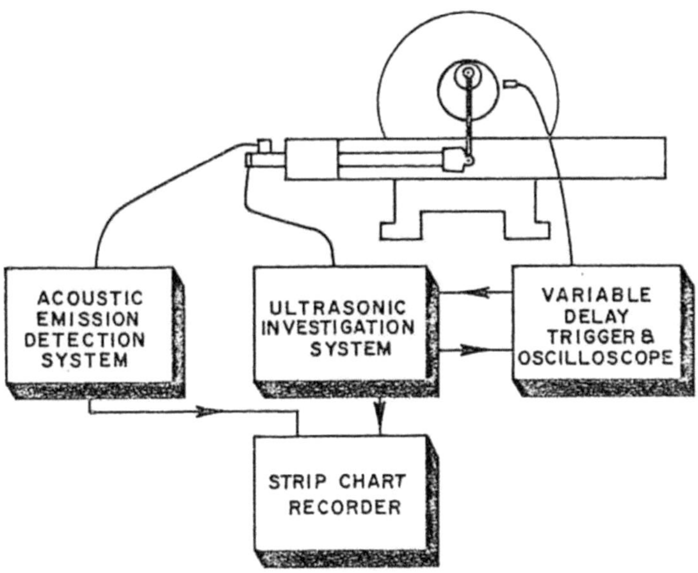

B

Figure 2.2.
(A) Schematic of experimental arrangement for ultrasonic attenuation detection of fatigue damage.
(B) schematic for ultrasonic and acoustic emission detection of fatigue damage.

presence and growth of fatigue cracks. This technique provides ample and early warning of impending failure and would therefore be of value in practical applications.

2.2. Composite materials

Composite materials can be arranged into three groups: dispersion strengthened, particle reinforced, and fiber reinforced. Composite materials are often both inhomogeneous and nonisotropic (orthotropic or anisotropic). Fiber-reinforced plastic (FRP) composites have desirable engineering properties such as low weight, high stiffness, and damage tolerance. Because of these properties, they have become attractive in aerospace and other modern structural applications. These applications have created a need to monitor their in-service performance nondestructively. Flaws and cracks in FRP, no matter how small, generate elastic waves as they extend suddenly under service loadings of intermittent or cyclic nature. These emitted elastic stress waves can be monitored to discover sudden damage or accumulated damage over the time period. Hence, AE technology is an ideal nondestructive testing technique for this purpose. AE technology, in the case of composite materials, has been applied to characterize fracture of fiber, matrix cracking, fiber-matrix debonding, matrix crazing, and gross delaminations. The methods used to process and to quantify the obtained AE data are primarily amplitude distribution, ring-down counting, and spectral analysis. However, the results obtained by all of the above methods have limited capabilities to characterize the source mechanisms involved by identifying the exact AE signatures associated with each of them. This is because the results obtained are not only dependent on the AE source mechanisms but also upon the elastic wave propagation characteristics of an anisotropic medium such as FRP, the dispersion and attenuation of AE stress wave in a medium, the distance between an AE source (either on the surface or in the bulk) and a transducer, and geometric effects including the resonances and antiresonances in the specimens under the study.

Kline [8] performed a study on the influence of attenuation on acoustic emission measurements during simulated AE events on metallic and polymeric media. The amplitude and rise time reductions were both greater for polymeric media (polymethyl methacrylate) than for aluminum. The basic features of AE waveforms are obscured with increasing sample thickness, principally due to preferential attenuation of the higher frequency components in the waveforms. Hamstad [9] pointed out that the performance of a fiber composite article is often dependent upon the local conditions throughout the composite. These local conditions include matrix crack density, fiber damage or fraying, matrix content, matrix cure, moisture content, residual stress field, and matrix to fiber bonding. In short, it would be of great

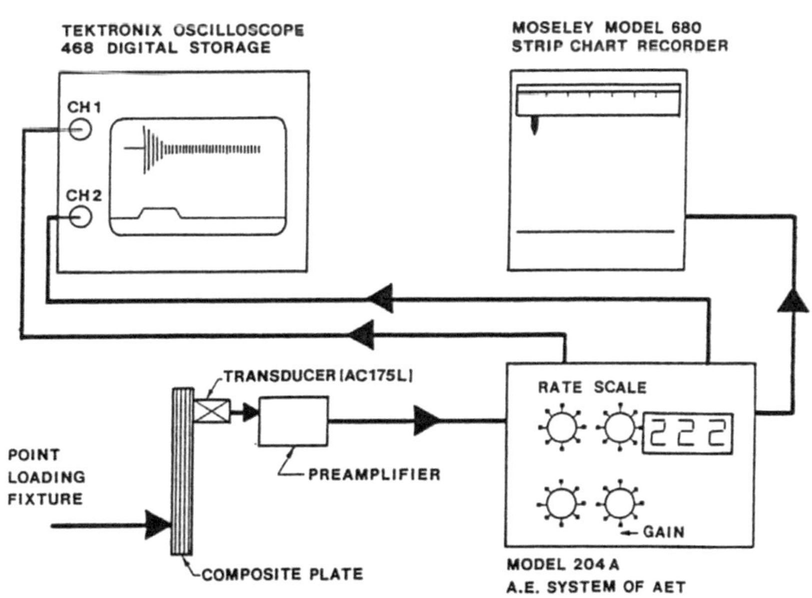

Figure 2.3.
Schematic drawing of the test apparatus.

help if the attenuation characteristics of composite plates were understood before their installation in-service components.

Such a study was performed by Joshi [10] on three carbon fiber-reinforced epoxy composites (CFRP). **Figure 2.3** shows the schematic drawing of the test apparatus. Plates of CFRP with different ply orientations were chosen for these attenuation studies. The first set was a unidirectional composite with a ply layup configuration of $[0_{22}]_T$. The second set of plates chosen had a ply layup configuration of $[(0/90)_{10}/0/(90/0)_{10}]_T$. The third set was a hybrid composite with a ply layup configuration of $[0/(0_3/\pm45)_8/0_3/(+45/0_3)_8/0]_T$. All of the plates were point loaded in steps until they cracked. The cumulative AE count above a constant threshold and at a constant gain was recorded for different orientations of a straight line from the AE transducer to the epicenter of the loading site with respect to the fiber positions in the plates.

Figure 2.4A shows the cumulative AE count versus loading steps for the case where the straight line joining the epicenter to the center of the transducer ran parallel to the fiber direction, and in **Figure 2.4B**, where it ran perpendicular to the fiber direction. The cumulative AE count measured in the perpendicular direction increased rapidly with fewer loading steps than those required to get the same count when measured in the parallel direction. In the second set of plates of the bidirectional composite, the number of loading steps to reach the same number of cumulative AE was close in number but still different when measured in two perpendicular directions (**Figure 2.4C** and **D**). In short, the attenuations of AE signals in the unidirectional composite are anisotropic and in the bidirectional relatively less anisotropic.

As shown in **Figure 2.4E**, for the third hybrid composite plate with complex ply layup configuration, the cumulative AE count before cracking of the plate was approximately the same for the same number of loading steps in all orientations. Thus, the cumulative AE count was an excellent indicator of localized damage, irrespective of the orientation of the straight line joining the epicenter of the damage source to the center of the transducer for fibers in both surface plies. The hybrid complex composite plate appeared more or less isotropic for the wave attenuation.

Proof testing with acoustic emission is very successful and is used to monitor tanks, pipes, and man-lift booms made of composite materials, using the standard procedures set forth by the Society of Plastic Industry and the American Society of Mechanical Engineers. Acceptance criteria are based primarily on the emission rate at constant proof loads. In FRP components, acoustic emission is often observed at loads lower than the previous maximum contrary to the Kaiser effect. This phenomenon is called the felicity effect. The Kaiser effect fails in viscoelastic situations such as the relaxation of the matrix in highly stressed composites. A flow of matrix at loads below the previous maximum can transfer stress to the fibers, causing them to break and emit AE. A felicity ratio of greater than 0.95 is recommended for applying AE to the inspection of FRP tanks and vessels. The Kaiser effect may be regarded in light of the felicity effect as a felicity ratio of 1.

2.3. Miscellaneous applications

The number of AE applications is increasing every day. AE techniques are used in monitoring weld quality, leak detection, stress corrosion cracking, hydrogen embrittlement, machinery conditions, and electric power plant structures. Acoustic emission techniques are also widely used in geotechnical applications, the electronics industry, the aerospace industry, biomedical engineering research, and even untended manufacturing. Chemical origins of AE are also being investigated. The

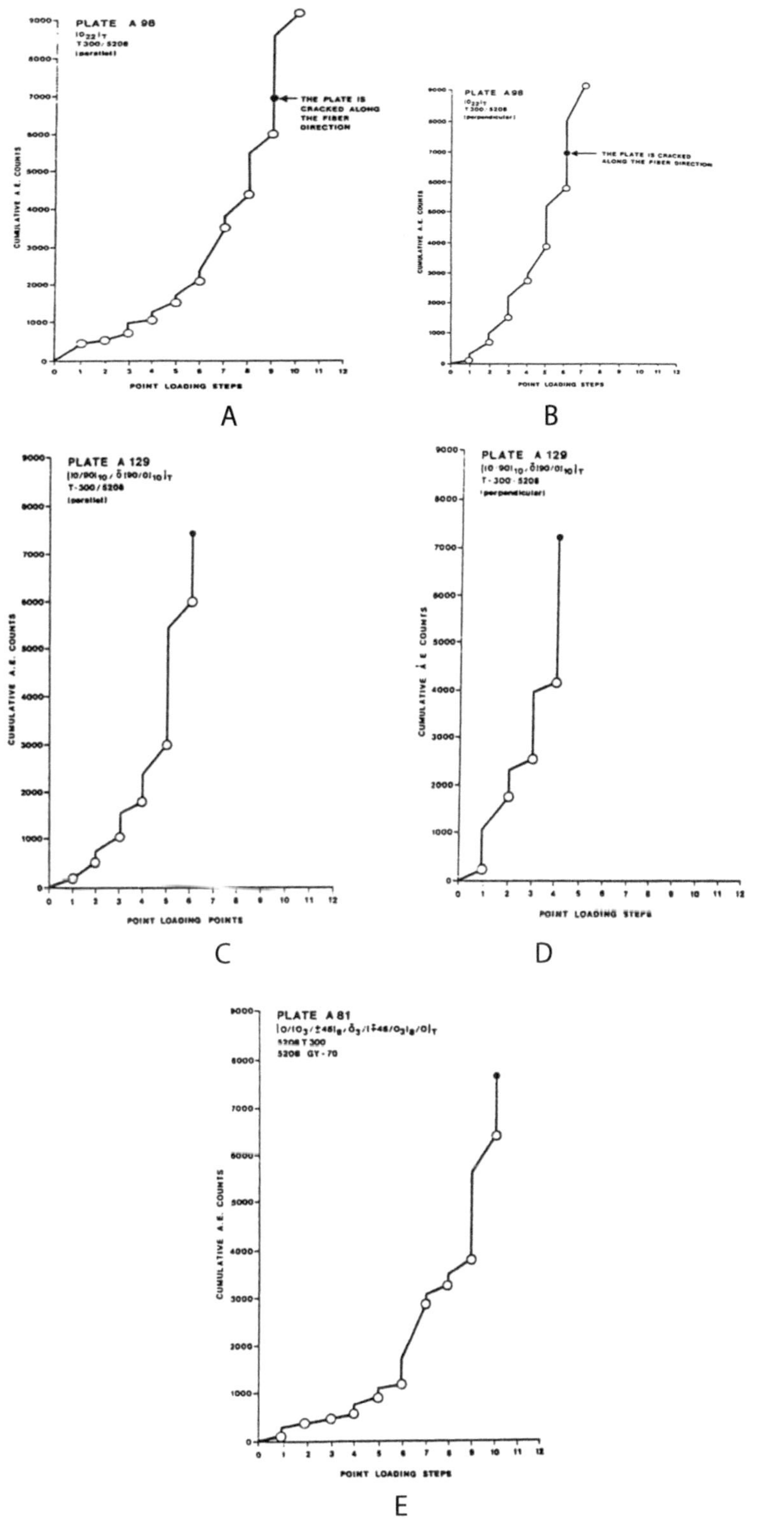

Figure 2.4.
(A) Unidirectional composite figure. (B) Unidirectional composite cumulative AE counts versus loading. Cumulative AE counts versus loading. AE signal detection parallel to fibers. AE signal detection perpendicular to fibers. (C) Bidirectional composite cumulative AE counts versus loading detection parallel to first fiberset. (D) Bidirectional composite cumulative AE counts versus loading AE detection parallel to second fiberset. (E) Hybrid composite cumulative AE counts versus loading AE count the same in all directions.

hydration of silica gel is a process in which a violent audible emission from gel granule fracture is accompanied by considerable emissions at ultrasonic frequencies. Many species of green plants emit acoustic signals during periods of water shortage. Emission frequencies extended to at least 300 kHz. The rate of drying of woody tissues can be monitored via the AE signals caused by fiber failures. A new application of AE was also recently discovered for in-situ performance monitoring of high-powered microwave tubes. The details of this novel research form the main topic of this book. The topic of application of AE to electrical and magnetic phenomena will be reviewed first because it indirectly led to the solution of certain problems related to high-powered microwave tubes.

2.4. Applications of acoustic emission to electromagnetics

Barkhausen, in 1917, discovered that voltages induced in an electrical coil encircling a ferromagnetic specimen subjected to a changing magnetic field produced a noise that could be heard through a speaker after proper amplification. This noise occurred even though the magnetization applied to the specimen was changed smoothly. From such experiments, he inferred that the magnetization in the specimen increases not in a strictly continuous way but by small abrupt discontinuous increments called Barkhausen jumps. During the magnetization process, these jumps are the means by which favorably oriented domains grow at the expense of less favorably oriented domains. In general, the magnetic behavior of a ferromagnetic material is determined by its microstructure and state of mechanical stress. Domain wall motion is influenced by the presence of inclusions, grain boundaries, and microstructure of the material. Hence, the Barkhausen effect can be expected to depend on many parameters, including stress and strain. Its possible use as a nondestructive testing technique related especially to residual stress measurement was pointed out by Scott [11]. Lord Jr., in 1974, detected AE pulses from domain wall jumps in nickel wire at frequencies around 100–200 kHz. The maximum Barkhausen noise in nickel occurs at the steepest portion of the hysteresis loop. The acoustic emissions also occur at the steepest part of the loop. The AE data were generally consistent with the Barkhausen effect. Larger Barkhausen effects are seen in cold-worked specimens and are accompanied by larger AE activity relative to annealed material. AE activity in annealed samples diminished considerably [12].

Beattie et al. [13] reported a spontaneous generation of acoustic signals during the switching of amorphous semiconductors. They suggested that AE activity could be used as a new tool for studying the switching process. Their experiments on bulk samples of chalcogenide glasses show two results. The first result is that in fast switching, and AE activity is associated with initiation of the switching process. The second result is that cracking in the glasses produces an easily detectable acoustic signal.

An electric arc within a superconducting winding may be induced by the breaking down of electrical insulation and by small metallic chip shorting conductors. The arc can cause serious damage to the magnet. Tsukamoto and Iwasa [14] used AE techniques to identify quench-causing sources and to detect the presence of a normalcy in superconducting magnets. Other applications listed by them included electric arc detection in pulsed superconducting magnets. They used differential sensors to discriminate acoustic signals from electromagnetic signals. A single-ended sensor responds to both acoustic and electromagnetic signals, while the reverse polarities of the two halves of a differential piezoelectric wafer subtract voltages of electrical noise and record AE activity only. Details of the differential sensor can be obtained from the ASNT Acoustic Emission Handbook.

Shen et al. [15] also used AE techniques to locate high-voltage breakdowns in large superconducting magnet systems. To ensure the safe discharge of superconducting magnet systems in case a quench or system malfunction occurs, the coils are designed and built to withstand high-discharge voltages. DC and transmission line techniques were unsuitable for finding the location of the dielectric breakdowns. They designed and developed an AE measurement system that would determine the locations of breakdowns in large coils using the triangulation method with AE sensors. The system measures the difference in time-of-arrival of transient waveforms caused by DC voltage discharges with an accuracy better than 5 cm. They used both AE measurements and the high-voltage testing system concurrently.

Acoustic signals in superconducting magnets are emitted principally by mechanical events such as conductor strain, conductor motion, frictional motion, and epoxy cracking. Because each event generates an acoustic emission signal, monitoring of both signal and magnet voltage makes it possible to identify and localize the source of an event responsible for a premature quench. Quench events in epoxy-impregnated superconducting magnets may be divided into three classes. Class 1: a quench preceded by a voltage spike and AE signal resulting from conductor motion. Class 2: a quench with AE signal but no voltage spike triggered by nonconductor-motion event such as epoxy cracking. Class 3: a quench with neither a voltage spike nor an AE signal resulting from pure Joule heating as the conductor reaches a critical current and enters into a current sharing regime.

Ige et al. [16] developed an acoustic emission system for monitoring high-performance superconducting dipoles and quadrupoles used in high-energy accelerators. They used four AE sensors, each one located at the top, middle, and bottom of the magnet, and the fourth, a guard sensor to detect extraneous noise, located on one of the current leads. In each run, the magnet was ramped up at a constant rate, typically 100 A per second, until it quenched at about 6000 A. They collected records on cumulative AE events and energy as well as current history. The data were then plotted separately for three AE sensors. They noticed that there was a sharp peak in the AE energy rate from the middle AE sensor. It thus suggested that this was the source of quench. It was independently verified later on using voltage data.

Similar results were obtained by Yoshida et al. [17] on large-scale coils of superconducting magnet systems. Uses of strain and displacement gages cannot cover the whole system in monitoring premonitory phenomena of a magnet system's failure, because these gages are mounted on points and therefore localized. On the other hand, acoustic emissions can be transmitted to sensors through structural materials without electrical noise. They used several sensors on large coil task (LCT) coil and the test module coil (TMC). They found correlation between voltage spikes and acoustic emission events excellent during single coil charging mode but poor during out of plane force mode. There were no indicative acoustical phenomena before a magnet quench or during normal zone generation.

Boczar [18] performed a frequency analysis of acoustic emission pulses generated during partial discharges in power devices. He discussed the disturbances, internal and external, that can accompany the measurement, using the AE method of partial discharges in industrial conditions. Interfering disturbances coming from the computer fan, the fluorescent lamp glow starter, the contact, the autotransformer, the electrical forcer, and the commutator engine, along with shifting the transducer along the surface of the object under study, were identified with their respective spectra. Methods were then suggested to eliminate their effects on the AE signal spectrum arising from partial discharges. He also performed a frequency analysis of AE signals for sliding and corona discharges to distinguish them from those of partial discharges. This analysis was needed because the AE signal frequencies of corona discharges (20–60 and 200–230 kHz) that occur on the equipment and phase conductors of overhead power lines often overlap the frequency spectrum of AE pulses generated by partial discharges. A high-pass filter with a bottom frequency of 60 kHz can eliminate the lower frequencies, and the higher frequency range may have some overlapping. Note, however, that the values for the amplitude and energy of AE pulses generated by sliding and corona discharges were over six times smaller than the corresponding values for AE pulses generated by the partial discharges that can occur in oil insulation setups [19].

In their study of AE signals generated during the uniaxial constant value compression of pristine cylindrical samples taken from the Kainda granite deposit (in Kyrgyzstan), Manzhikov et al. [20] discovered electromagnetic emissions (EME), which they attributed to charge redistribution caused by dislocation movements typical of most of the minerals (consisting of dielectrics with ionic bonds) forming the terrestrial crust.

Sakoda et al.[21] mounted AE sensors on the grounded outside surface of the tank of an oil-immersed pole transformer in order to study elastic waves generated by a simulated partial discharge or a corona discharge in their laboratory experiment. In the first step, there were no windings or magnetic core in the oil-filled tank. They then repeated the experiment with the windings and core included in the tank to figure out the modification of the elastic Lamb waves in the thin tank wall due to the presence of the core and windings. The results showed that the windings influenced the characteristics of the frequency spectra of the elastic waves, and that most of the detected AE signals were for the Lamb waves having frequency components greater than 50 kHz. It was observed experimentally that these frequency components were mainly caused by the elastic waves passing through the surface of insulating paper on winding layers and oil with repeating transmission, refraction, and reflection. The Institute of Electrical and Electronics Engineers IEEE has produced an IEEE-Trial-Use Guide (IEEE Std C57.127-2000) for the detection of Acoustic Emissions from Partial Discharges in Oil-immersed Power Transformers (IEEE Std C57.127-2000).

References

[1] Miller RK, McIntire P, editors. Nondestructive testing handbook. In: Acoustic Emission Testing. Vol. 5. Ohio, USA: American Society for Nondestructive Testing; 1987

[2] Pollock AA. Acoustic emission. In: Stephens RWB, Leventhall HG, editors. Acoustics and Vibration Progress. Vol. I. London, UK: Chapman and Hall; 1974. pp. 51-83

[3] Dunegan H, Harris D. Acoustic emission-a new nondestructive testing tool. Ultrasonics. 1969;7(3):160-166

[4] Joshi NR. Ultrasonic detection of fatigue damage [Ph.D. thesis]. Baltimore, MD: The Johns Hopkins University; 1973

[5] Joshi NR, Green RE Jr. Ultrasonic detection of fatigue damage. Engineering Fracture Mechanics. 1972;**4**:577-583

[6] Green RE Jr, Duke JC Jr. Ultrasonic and acoustic emission detection of fatigue damage. In: International Advances in Nondestructive Testing. Vol. VI. New York, NY: Gordon and Breach Science Publishers; 1979

[7] Harris DO, Dunegan HL, Tetelman AS. Prediction of fatigue lifetime by combined fracture mechanics and acoustic emission techniques. In: Proceedings of the Air Force Conference on Fatigue and Fracture of Aircraft Structures and Materials AFFDL, TR 70-144; 1969

[8] Kline RA. The influence of attenuation on acoustic emission measurements. IEEE Ultrasonic Symposium Proceedings. 1984;**2**:930-932

[9] Hamstad MA. Local characterization of Fiber composites by acoustic emission. In: Critical Review: Techniques for the Characterization of Composite Materials, Cambridge, MA. Boston, MA: Office of Naval Research, Eastern/Central Regional Office; 1981

[10] Joshi NR. Acoustic emission studies of fiber reinforced composites. In: Proceedings of 15th Symposium on Nondestructive Evaluation. San Antonio, Texas: SWRI; 1985. pp. 320-330

[11] Scott IG. Basic Acoustic Emission. New York, NY: Gordon and Breach Science Publishers; 1991

[12] Lord AE. Acoustic emission. In: Mason WP, Thurston RN, editors. Physical Acoustics. Vol. XI. Cambridge, MA: Elsevier. 1975. pp. 289-353

[13] Beattie AG, Johnson RT Jr, Quinn RK. Spontaneous Generation of Acoustic Signals during Switching of Amorphous Semiconductors. Albuquerque, NM: Sandia Laboratories; 1974

[14] Tsukamoto O, Iwasa Y. Acoustic emission diagnostic & monitoring techniques for superconducting magnets. Advances in Cryogenic Engineering. 1986;**31**:259-268

[15] Shen SS, Wilson CT, Luton JN. Acoustic emission measurements for locating high-voltage break downs in large superconducting magnet systems. Advances in Cryogenic Engineering. 1986;**31**:285-292

[16] Ige OO, Iwasa Y, Fujita H. Acoustic emission instrumentation for a superconducting dipole. Advances in Cryogenic Engineering. New York: Gordon & Breach Publishers, Inc.; 1979;**31**:303-308

[17] Yoshida K, Nishi MF, Tsuji H, Hattori Y, Shimamoto S, Tsukamoto O. Acoustic emission measurement on large scale coils at JAERI. Advances in Cryogenic Engineering. 1986;**31**:277-284

[18] Boczar T. Identification of fundamental forms of partial

discharges based on the results of frequency analysis of their acoustic emission. Journal of Acoustic Emission. 1999;**17**(3-4):S7-S12

[19] Boczar T. Results of Frequency Analysis of AE Pulses Generated by Corona Discharges. Poland: Technical University of Opole; 2004. www.pu.if.ua/inst.phys_che/start/pcss/vol5/0504-28.pdf

[20] Manzhikov BT, Bogomolov LM, Sychev VN. Structure of Pulses of Acoustic Emission and Electromagnetic Emission from Rocks during Uniaxial Compression. http://tiger.gdirc.ru/1201/1201_en/emi.html

[21] Sakoda T, Nieda H, Ando K. Characteristics of elastic waves caused by Corona discharges in an oil-immersed pole transformers. IEEE Transactions on Dielectrics and Electrical Insulation. 2001;**8**(2):276-283

Chapter 3

Nondestructive Evaluation of Microwave Tubes

3.1. What are microwaves?

Microwaves are electromagnetic waves that have many similarities to visible light and obey the same laws of optics. They can travel in straight lines over great distances, reflect from objects in their path, and allow measurements of the reflected or transmitted waves to define position, speed, diffraction, refraction, energy absorbed, and other characteristics. Microwaves can be focused by lenses or by curved mirrors, and they travel in rays like light. They reflect, refract, diffract, and experience interference and scattering. However, they differ from light in frequencies and wavelengths. **Figure 3.1** shows the relative position of microwaves between radio waves and infrared radiation in the electromagnetic spectrum. The figure also shows the values of frequencies and wavelengths of different regions of the electromagnetic spectrum as related through the equation $v = f\lambda$, where v = velocity of waves, f = frequency, and λ = wavelength. A typical microwave wavelength would be 25 mm (1 inch), while that of television would be 0.4–5.5 m (1.3–18 ft) and that of AM radio around 300 m (1000 ft). Sometimes the term microwave is used rather ambiguously. However, signals at frequencies from 1 GHz (10^9 cycles/s) to at least 100 GHz are considered to be microwaves. Microwaves propagate through free space as linearly or circularly polarized, sinusoidally varying, plane electromagnetic waves at the speed of light, equal to 2.988 × 108 m/s.

Microwaves are used widely in modern technology. One major field is television. The microwave frequency range is used for transcontinental transmission using complex transmitter, relay, and receiver networks throughout the country. Local stations convert the signals to lower frequencies for transmission to individual sets. On a local level, microwaves are used to send signals from studios to transmitter locations. Microwaves are used in national and local security applications, such as early warning radar, missile guidance systems, and Doppler radars, to detect and control the speed of vehicles. Microwave communication networks and relay stations are used commercially for routine multichannel communication transmissions, both long distance and local. With microwave technology, air and sea navigation is much more reliable now than in the past. On the consumer level, microwave ovens have become very widespread. One can add to this list the industrial applications of microwaves, such as microwave heating, drying, curing, etc. Chemical and biomedical applications are also steadily increasing.

3.2. Types of microwave devices

Microwave tubes are widely used in radar transmitter applications. Microwave radar transmitter tubes are of two types, oscillators and amplifiers. Self-excited oscillators, as typified by magnetrons, and various semiconductor bulk-effect devices, such as Gunn effect devices, impact avalanche and transit-time devices (IMPATTs), and limited space charge accumulation (LSA) devices, are suitable for many radar applications. Amplifiers, which include cross-field amplifiers (CFAs),

Type of Radiation (Regions overlap)	Frequency in Hz	Wavelength
Radio Waves (FM and TV)	10^3 to 10^9	10,000 meters to 1 meter
Microwaves	10^9 to 10^{12}	10 cm to 1 mm
Infrared	10^{12} to 10^{14}	1 mm to 1 μm
Visible Light	10^{14} to 10^{15}	750 nm to 400 nm
Ultraviolet	10^{15} to 10^{17}	400 nm to 10 nm
X rays	10^{17} to 10^{19}	10 nm to 1 nm
Gamma rays	10^{19} to 10^{25}	1 pm to 0.001 pm
Note: 1 pm = 10^{-12} meter Picometer	1nm = 10^{-9} meter Nanometer	1 μm = 10^{-6} meter Micrometer

Figure 3.1.
Position of microwaves in electromagnetic spectrum.

klystrons, traveling-wave tubes (TWTs), and transistor amplifiers, are useful where the amplification of low-power signals is required, as in a coherent MTI system, a pulse compression radar, or a pulse-Doppler system. Many radar systems operate in a pulsed mode and require transmitter devices capable of considerable peak and average power. While there are exceptions, applications requiring high output powers usually use thermionic vacuum tubes for the transmitting devices.

High-power microwave (HPM) tubes may be divided into two categories: crossed-field devices, characterized by orthogonal electric field and magnetic fields [1, 2], and linear-beam devices, which have a continuous electron beam traversing an interaction region [3]. Crossed-field devices are typified by magnetrons and CFAs, while linear-beam devices include cavity klystrons, traveling-wave tubes, twystrons, and extended-interaction oscillators. Each of these devices has its own peculiarities concerning arcing, voltage-current relationship at the terminals, RF performance, and the effects of terminal voltages on tenability and device stability.

Out of all these devices, three types of transmitting devices have been dominant in radar for many years. They are magnetrons, klystrons, and traveling-wave tubes. They all are based on vacuum-tube technology. Magnetrons are rugged and suitable for mobile equipment. Klystron amplifiers offer the system designers a higher peak power capacity than magnetrons. Since they are amplifiers rather than oscillators, they can provide highly stable output pulses, controlled in phase and frequency. On the other hand, klystrons are comparatively bulky, expensive, and prone to mechanical damage. Traveling-wave tubes, which are close relatives to klystrons, are also amplifiers. They have relatively large bandwidths. HPM tube systems are used in Navy radars and communication systems. These tubes are primary cost drivers, as can be seen from expenditures of millions of dollars on replacing malfunctioning tubes in the fleet. Sometimes it is difficult to make decisions about replacement of the expensive tubes in the absence of tools to examine their conditions.

3.3. Need for nondestructive evaluation (NDE) of microwave tubes

As mentioned before, microwave tubes are primary cost drivers for many high-power radars and communication systems. It is not unusual for high-performance microwave tubes to cost in the range of 20,000–30,000 dollars. Millimeter wave tubes can cost much more. Since a tube's operating life may only be a few thousand hours, an increase in tube life can significantly reduce annual system operating costs. Incompatibility between a tube and its system can also seriously impact tube life. Although efforts have been made to achieve compatibility between tubes and their systems, this has rarely provided the protection required to achieve long tube life. When a costly tube fails, it is generally impossible to determine the sequence of events that contributed to its failure. This makes it difficult to correct the tube's design or to improve system interfacing. Written reports, returned with failed tubes, seldom provide the kind of information needed to clearly establish the cause of failure. This state of affairs clearly indicates that techniques for in situ performance monitoring of HPM radar tubes are much needed.

3.4. Requirements of nondestructive evaluation of microwave tubes

The objectives of the NDE project were to accomplish the following:

1. Determine the measurable characteristics that correlate with tube status and reliability.

2. Develop sensors for monitoring these characteristics.

3. Integrate these sensors with telemetry to obtain a nonintrusive package.

4. Develop automated data collection and analysis tools.

5. The requirements guidance are as follows:

    a. Insert the technology into in-service systems. The NDE package should provide upgrades of in-service systems with new, non-intrusive monitoring technology.

    b. Insert technologies that reduce the cost of ownership of systems. The NDE package should provide advice on judicious and timely replacement of high-value microwave power tubes, that is, condition-based maintenance.

The microwave power vacuum tube types selected for the project were:

1. Linear beam family

    a. Multicavity reflex klystron

    b. Traveling-wave tube helical coupled cavity

2. Crossed-field family

    a. Magnetron

    b. Crossed-field amplifier

The common characteristics of these tubes were identified as:

1. High internal vacuum (10^{-7}–10^{-8} Torr)

2. High electrode voltages (>10 kV)

3. High heat dissipation (100 W–10 kW)
4. Heated cathode
5. Expensive

Performance monitoring for a HPM tube in a radar system could be a complicated procedure as can be seen from the following partial list of currently monitored functions on a radar system having a klystron microwave tube.

1. Filament voltage
2. Filament current
3. Focus coil current
4. Focus coil voltage
5. Vacuum-ion voltage
6. Vacuum-ion current
7. Waveguide pressure
8. Waveguide VSWR/arc detection
9. Klystron temperature
10. Klystron airflow
11. Klystron oil level
12. Klystron oil flow
13. Klystron oil temperature
14. Klystron cathode current
15. Transmitter overvoltage
16. Transmitter overcurrent
17. Cabinet temperature
18. Cabinet airflow
19. Low transmitter output power
20. Modulator inverse current
21. Low voltage power supply (LVPS) voltages

22. Body current

23. Collector current

3.5. Currently used nondestructive evaluation methods

3.5.1 Thermal inspection

Generally, thermal inspection measurements fall into two categories. First is thermometry, which is related to measurement of temperature, and second is themography, which maps contours of equal temperature (isotherms) over a surface. Sensing devices operate in two modes, contact mode and noncontact mode. Contact sensors such as thermocouples, bolometers, thermistors, paints, and inks are either applied to or touch the surface of the test object. Noncontact methods rely on remote sensing of infrared radiation. Pulses of heat energy, from a source, are directed at the component under test. It is usual, but not essential, to direct the incident energy onto one surface of a component and observe the effects at the opposite surface after conduction through materials. Flaws and irregularities in structures will affect the amount of conduction in their vicinity. In some cases, the test object is heated by its internal operation and the heat radiated from its surface can be detected with an infrared camera and used to map the uneven surface temperature contours or points of excessive high temperatures. In another version of thermal pulse video thermography, a high-intensity heat source is used to send pulses of infrared energy into the material. The surface is then scanned by an infrared thermal imager with a video output. Digital image processing that provides image enhancement is also possible.

In the current work, an Inframetrics IRTV-45L (FLIR WIR 8- to 12-μm) camera was used to examine hot spots on a traveling-wave tube (HP 495A microwave amplifier operating at 8.5 GHz). **Figure 3.2A** shows the setup and thermal images.

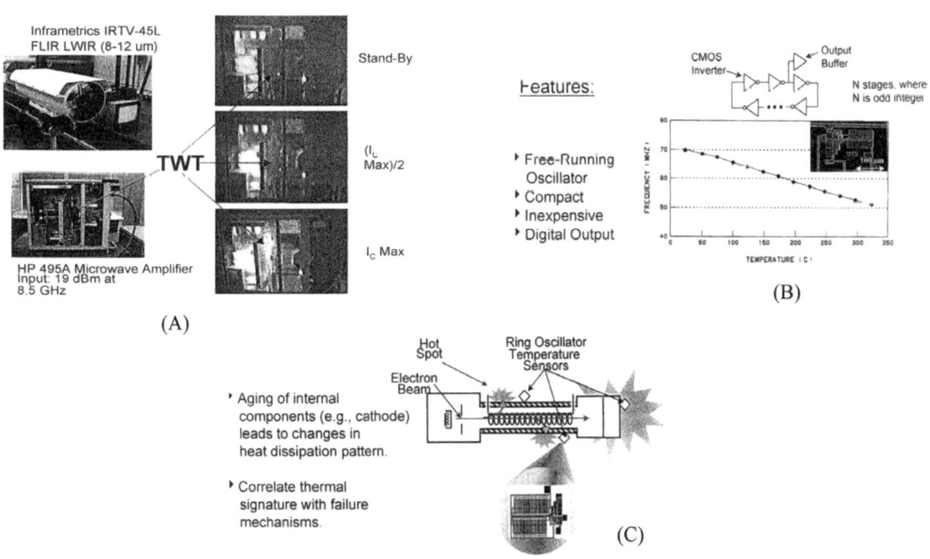

Figure 3.2.
(A) Infrared camera to detect hot spots on TWTA tube. (B) Frequency versus temperature for CMOS ring oscillator. (C) Thermal mapping with CMOS ring oscillators on TWT.

Next, a CMOS ring oscillator was used as the temperature sensor [4]. The oscillator is free running, compact, and inexpensive and offers a digital output. **Figure 3.2B** shows a schematic of the sensor and its output frequency versus temperature.

Figure 3.2C shows a schematic diagram of thermal mapping of a traveling-wave tube. The TWT undergoes an aging process during its service use. Aging of its internal components such as the cathode leads to changes in the heat dissipation pattern. It has been proposed that a tube's thermal signature could be correlated with its failure mechanisms.

3.5.2 PIN diode sensors

Collector depression is the process of applying a negative potential (on the tube body or "ground") to the collector of a TWT or klystron to reduce the electron beam velocity as the electrons enter the collector. This reduces the kinetic energy of the electron beam, causing less energy to be converted into X-rays and heat when the beam impinges on the internal surfaces of the collector. The rate of emission of these X-rays could be used to estimate the tube aging, as the amount and distribution of these X-rays changes with tube aging. PIN diode sensors are used here because they can measure X-ray radiation levels and can localize their sources [5]. They could serve the double purpose of indicators of tube aging as well as monitoring X-ray emissions to secure personnel safety. **Figure 3.3** shows a schematic of an X-ray detection system with PIN diode sensors on a TWT.

3.5.3 Optical methods

For voltage hold-off in vacuum devices such as X-ray tubes, klystrons, or magnetrons, it is important to realize that each device presents its own problems and that no all-embracing rules can be established. The specifics of vacuum arcs depend upon the region near the cathode, where the cathode spot plasma processes of heat and mass transfer are intensive and the charge carriers are generated. A nonintrusive method could be used to monitor materials evaporated from the cathode, which get deposited on the interior walls of the vacuum tubes. Changes over time in the thickness of the deposited layer could be used to predict the remaining cathode life. It has been suggested that optical methods, comparing reflections of light from

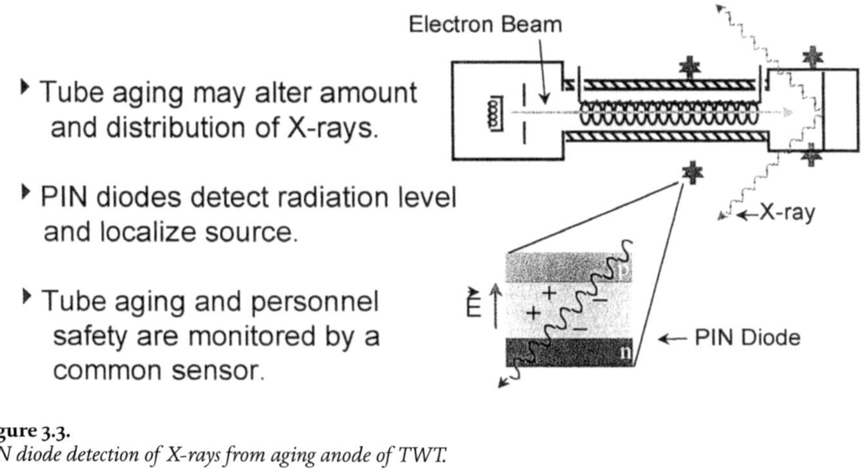

Figure 3.3.
PIN diode detection of X-rays from aging anode of TWT.

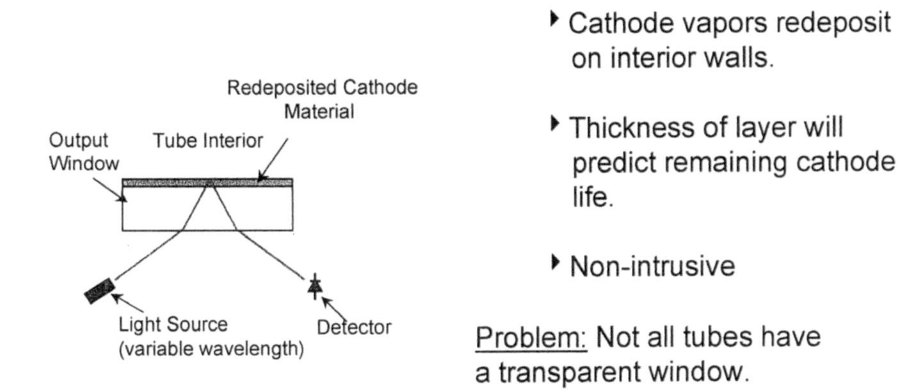

Figure 3.4.
Ellipsometric measurement of deposited vapors from aging cathode.

clean interior walls to those with deposited layers, could be used to estimate the remaining cathode life. One drawback of this NDE method is that it cannot be used for tubes that do not have transparent windows to shoot the light beam. **Figure 3.4** shows a schematic of the optical method.

3.5.4 Giant magnetoresistance (GMR) sensor

One way to study the performance of HPM tube is to monitor cathode current pulses, distinguishing bad current pulses from normal ones. This could be achieved with a current sensor [6]. **Figure 3.5A** shows the operation of a GMR sensor. **Figure 3.5B** illustrates the advantages of a GMR sensor over a conventional current sensor. The conventional current sensor is a transformer whose secondary winding puts out 1 V per ampere of cathode current during pulsing of the tube. The advantages of GMR sensors are their compactness, low relative cost, and noncontact operation.

3.5.5 Standard electrical sensor method

Electrical sensors monitoring cathode current, cathode voltage, RF pulse and RF power, and outputs of microwave tubes can be connected to different channels of an advanced digital oscilloscope to view and store in its memory. In this way, one can compare the shapes and characteristics of anomalous pulses with those of a standard one.

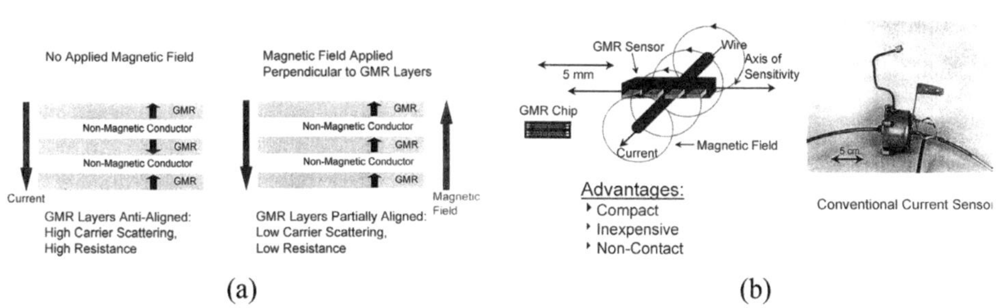

Figure 3.5.
(A) Schematic of operation of GMR sensor. (B) Nonintrusive current measurement using GMR sensor.

3.6. Concept of microprocessor-based monitoring system

As mentioned before, written reports, returned with failed HPM radar tubes, seldom provide the kind of the information needed to clearly establish the cause of their failure. So Watson and Troy [7] envisioned a microprocessor-based performance monitoring system that could provide capabilities for monitoring tube performance, providing tube protection, and recording a comprehensive tube failure history. The performance monitoring concept, as depicted schematically in **Figure 3.6A**, shows a typical situation applied to a high-value, crossed-field amplifier. **Figure 3.6B** shows a conceptual block diagram of such a monitor, which utilizes a microprocessor to control a series of data acquisition components. In this configuration, the microprocessor analyzes the measured data to sense faults. It then alerts the system and stores the analyzed data. The long-term goal for their program was to develop a tube performance monitoring system with the following characteristics:

1. Wide applicability

2. Software controlled

3. Self-calibrating

4. Detailed resolution (50 ns)

5. Multiple sensors

6. Automatic fault sensing

A brief narration on each of the desired characteristics of their performance monitoring system is given in **Figure 3.6**.

3.6.1 Wide applicability/software controlled

The wide applicability requirement almost automatically dictates the use of software control. The wide applicability capability is very important because it allows the performance monitor to be used with a variety of different tube systems.

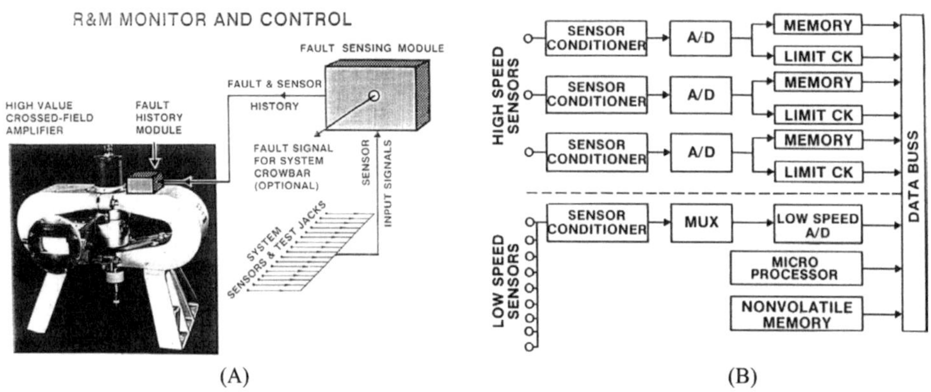

Figure 3.6.
(A) Performance monitoring concept for microwave device. (B) Conceptual block diagram of performance monitor.

Software control ensures a high degree of flexibility because it allows algorithms to be tailored to components, systems, and operational circumstances. In addition, it provides growth potential, since improved algorithms can be implemented as tube protection knowledge increases.

3.6.2 Self-calibrating

The self-calibrating philosophy is based on the assumption that there will exist a reasonable time period during which the tube operates normally and that the performance during this time can be used as a measure of "normal performance." The process of self-calibration can be implemented by computing an initial mean and variance for each data set to be evaluated. The data set for each time period is called a resolution bin. Each sample can then be compared to its stored mean value. The use of variance, or other equivalent variability measurements, allows the setting of a detection threshold to detect faults with very low false alarm rates. The mean and variance values can be continually updated during monitor operation to allow for normal system variability. In addition, multisample and multipulse fault criteria can reduce the false alarm rates to very low levels. Self-calibration thus eliminates the difficult task of establishing and maintaining an absolute calibration of system and monitoring equipment.

3.6.3 Detailed resolution

Detailed resolution is required to provide usage of all of the available sensor information that bears on a tube's performance. In practice, HPM radar vacuum tubes under stress often exhibit a slight distortion of the tops of the cathode current pulses, along with leading edge jitter, and experience changes in modulator voltage and load current. Most of these small effects cannot be detected using a low-resolution monitoring system. As a general rule, it is desirable to have a sufficient number of samples to allow reconstruction of the actual pulse shape with reasonably good fidelity, characteristic of a pulse being viewed on a high-end digital oscilloscope. Clearly, a typical radar current pulse cannot be adequately reproduced using five or six time samples. It requires at least 15–20 samples to adequately reproduce the structure of the pulse. Thus, a high-speed data sampling rate of 20 million samples per second should be chosen for the performance monitoring system since the length of a typical pulse is about 1 μs. For longer pulses, fine structure usually becomes less important. While a high analog-to-digital conversion rate clearly increases resolution, it could introduce buffering and throughput problems.

3.6.4 Multiple sensors

Multiple sensors provide the opportunity to implement algorithms that sample several signals simultaneously. This allows an in-depth multisensor analysis of abnormal events.

Furthermore, it allows the monitoring system to ignore extraneous events that may affect a single sensor but do not actually affect tube operation. Substantially increased information concerning a fault, and its causes, can be derived from concurrent multisensor analysis.

3.6.5 Automatic fault sensing

Automatic fault sensing capability is an essential feature of a stand-alone performance monitor. The detection of a fault is the first step in alerting the system and in preparing it to store data on the fault for later analysis. Automatic fault sensing provides a very efficient means for selecting meaningful experimental data for analysis. For laboratory experiments, automatic fault sensing is very preferable over manually stopping the measurement system for the recording and analysis of data, on the basis of indications viewed on an oscilloscope. Furthermore, the use of automatic fault sensing in the breadboard system would provide an opportunity to try out various fault sensing concepts and to evaluate their effectiveness in an operating environment.

3.7. Final comments

This completes the summary of the basic concepts of a microprocessor-based, automated, nondestructive evaluation system for high-power microwave tubes. The next chapter describes the implementation of a breadboard prototype of such a system.

References

[1] Lau YY. Theory of cross field devices and a comparative study of other radiation sources. NRL Memorandum Report, 6029. 21 July 1987. http://www.dtic.mil/dtic/tr/fulltext/u2/a183389.pdf

[2] Chodorow M, Susskind C. Fundamentals of Microwave Electronics. New York, NY: McGraw Hill Book Co.; 1964. pp. 241-244

[3] Liao SY. Microwave Devices and Circuits. Englewood Cliffs, NJ: Prentice Hall, Inc.; 1980. pp. 177-255

[4] Suman S, Singh BP. Design of temperature sensor using ring oscillator. International Journal of Scientific and Engineering Research. May 2012;**3**(5)

[5] Owen RL, Holton JM, Schulze-Briese C, Garman EF. Determination of X-ray flux using silicon pin diodes. Journal of Synchrotron Radiation. 2009;**16**:143-151. Published Online 2009 Feb 25. DOI: 10.1107/S0909049508040429

[6] Ouyang Y, He J, Hu j, Wang SX. A current sensor based on the giant magnetoresistance effect: Design and potential smart grid applications. Switzerland: Sensors, Basel; 2012;**12**(11):15520-15541. Published Online 2012 Nov 9. DOI: 10.3390/s121115438

[7] Watson WH, Troy JW. Computer-controlled performance monitor for microwave tubes. In: NOSC* Technical Document 1070. San Diego, CA: Naval Ocean Systems Center; 1987. *This organization is now called Space and Naval Warfare Systems Center (SSC Pacific)

Chapter 4

Development of the Performance Monitoring System

4.1. System design

A breadboard version of a computer-based performance monitoring system was assembled to check its feasibility. An initial monitoring system for high-value microwave tubes was developed during the 1970s and 1980s by Watson and Troy [1, 2]. To assemble the breadboard system, components were chosen depending on their availability during the 1970s and 1980s. The components available at the time of their development effort had limited data collection and analysis rates. The equipment purchased included a 13-slot Model 8013A CAMAC enclosure fitted with an integral power supply, three 20-MHz analog-to-digital (A/D) converters, one low-speed multiplexed A/D converter, and a CAMAC controller with IEEE-488 (GPIB) interface capability, all from LeCroy Corporation. The monitor design included a Hewlett-Packard HP-1000 computer to simulate the microprocessor functions. For the breadboard system, this general-purpose computer provided important capabilities for program generation and data analysis. Communication between the CAMAC crate and the HP-1000 computer was via the GPIB bus. The passive crate controller limited the data transfer rates but did not seriously impact the primary goals of the program. **Figure 4.1** shows a simplified block diagram of this breadboard system. The basic A/D modules were mounted in the CAMAC crate, which in turn was connected to the HP-1000 computer via the GPIB bus.

4.1.1 Sensors

The monitoring system was designed to accommodate three high-speed channels and up to eight low-speed channels. Since the system was designed to be self-calibrating, it was only necessary to ensure that signals from all of the sensors fell within the allowable limits of the A/D converters. The high- and low-speed A/D converters allowed maximum inputs of 512 mV, and both incorporated input-level offset adjustments. The low-speed A/D converter also used polarity switches for each channel. The combination of these features provided the ability to handle positive, negative, and bipolar inputs. The high-speed A/D converters also included built-in attenuators to facilitate adjustment of input levels. In almost all cases, the sensor signal levels were within the acceptable limits for all of the A/D converters. The attenuation in the sensor input lines solved most of the level problems with the sole exception of a temperature sensor, the thermocouple that had a very low output level. Hence, the thermocouple sensor channel required 40 dB of amplification to bring it up to acceptable levels.

4.1.2 System control

For field applications of a performance monitoring system, the trigger pulses for operating the radar tube must also initiate the data acquisition process to

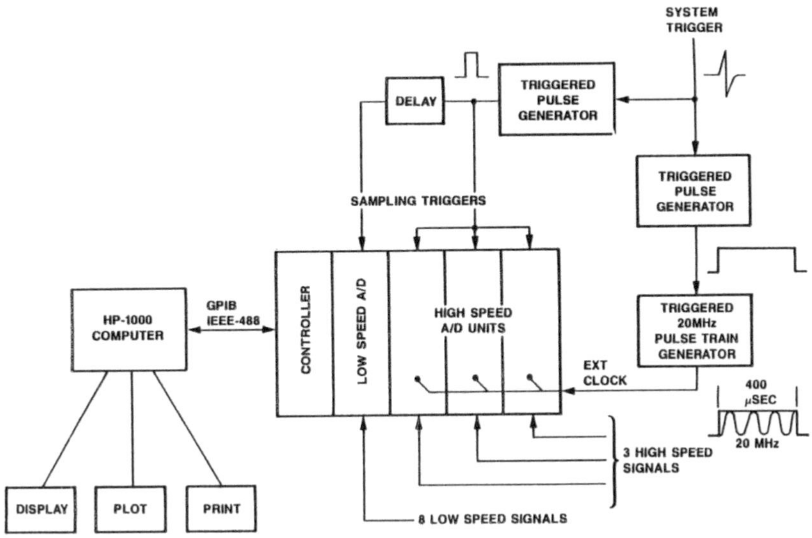

Figure 4.1.
Simplified block diagram of the breadboard system (after Watson and Troy [2]).

ensure synchronization. This applies to both pulsed and nonpulsed operations, since there is always a sequence of system data events that must be synchronized for effective signal processing. In the breadboard system, the HP-1000 computer activated all of the data acquisition modules and started the data acquisition process. Each high-speed A/D converter contained a first-in-first-out (FIFO) memory of 1024 8-bit words, with provisions to select a specified number of pretrigger samples.

In operation, the high-speed A/D converter started data acquisition when it received the system trigger. The A/D converter signaled the CAMAC crate controller when its buffer memory was full. This information was then provided to the HP-1000 computer, which stopped the data acquisition process and initiated the transfer of data via the GPIB interface. Sixty-four 8-bit samples were transferred from each high-speed unit on every system pulse. All of these data had to be transferred during the interpulse period. For example, if the performance monitor was used with a radar system operating at a repetition rate of 200 pps, the interpulse period would be 5 ms. Therefore, during this 5 ms, all acquired data had to be transferred, fault tests performed, mean and variance calculated, the system reinitialized, and the data acquisition cycle restarted. If all these functions could not be accomplished during the interpulse period, the performance monitor would go out of synchronization and fail to operate properly. Low-speed data were not used to determine whether a fault had occurred. During the measurement process, all low-speed data were stored in a 32 K memory module located in the CAMAC crate. After the data series was completed with the detected faults, the low-speed data were transferred to the HP-1000 and used with all of the other data in the fault analysis.

4.1.3 Data transfer

The rate of transfer of data from the high-speed A/D converters to the HP-1000 computer was the limiting factor in determining the overall system speed. The data transfer rate was primarily limited by the time required to execute the HP-1000 software and the additional time utilized by the HP-1000 GPIB driver. The data transfer software was written in assembly language to increase the speed. Using high-speed techniques, it was possible to reduce the transfer time for 64 data samples from a single high-speed A/D converter to approximately 1 ms. Thus, 3 ms was required to transfer data from all three high-speed A/D converters. An additional 2 ms was required to perform out the data analysis. These time restrictions therefore allowed a maximum repetition rate of 200 pulses per second of the pulsing radar tube.

4.1.4 Data analysis

After the transfer of high-speed A/D data to the HP-1000 computer, a very short series of calculations and tests were conducted. Part of this computation was concerned with determining whether a fault had occurred. The balance of the analysis was concerned with determining expected data limits for use in later fault analysis. The fault tests were not performed for the first 16 pulses to determine initial expected values. Several comprehensive techniques were used to compute these expected values, but the method finally selected was to determine the initial maxima and minima bin values by testing an initial series of 16 pulses. These expected values were continually updated with each subsequent pulse received. This technique worked well and was very fast. The fault tests were conducted by testing each data sample against the maximum and minimum values. A major fault was considered to have occurred when five or more bin faults occurred on two successive pulses. These fault tests were performed on the RF data. After the occurrence of a major fault, 16 additional pulses were measured to provide 32 pulses for use in subsequent analysis. At this time, the low-speed sensor data, stored in the 32 K memory, were transferred to the HP-1000 computer and stored with the high-speed A/D data for later analysis.

4.1.5 Display and plot capability

A variety of printing, display, and plotting capabilities were available to support the goals of this analysis. Once the A/D data had been transferred to the HP-1000 computer, the operator could print out hard copies of all of the data. In addition, the operator could select one or more sets of pulse data for display on the HP-1000 display terminal. These data displays made it possible to rapidly identify data of principal interest. After determining which data to display, the operator could use a separate color plotting capability to generate color plots of the selected data. Another useful printout available was the fault matrix shown in **Figure 4.2**, plotted using data stored during the fault testing.

Figure 4.2 shows 40 time bins for each of the 32 stored pulses. All matrix bins were set to zero (blank) if no fault occurred, and set to one (black) if a fault occurred. The experimental process continued the storing of 16 additional pulses after the occurrence of a major fault. For the example shown, one can see that initial bin faults occurred at pulse 15 and continued into pulse 16. After measuring and recording the 16 additional pulses, the experiment was automatically stopped. From **Figure 4.2**, it is evident that the tube initially faulted at the trailing edge of the 15th pulse and continued to fault in this same region. This type of display helped define where faults had occurred.

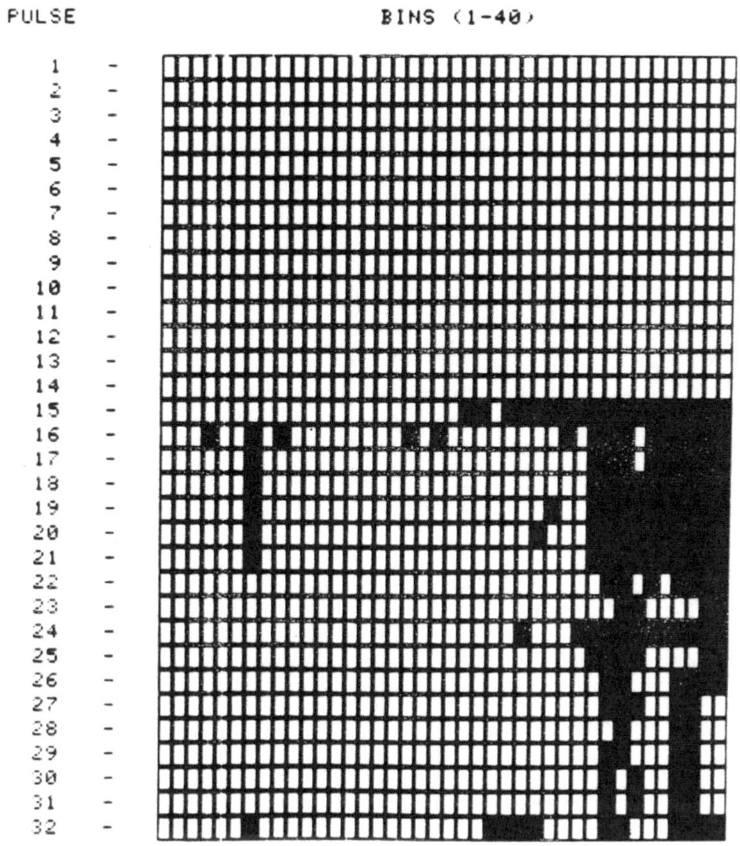

Figure 4.2.
Fault matrix printout (high-speed sensor 1 used) bin faults—solid black color (after Watson and Troy [2]).

4.2. Deployment of the performance monitoring system for magnetrons

4.2.1 Description of experimental apparatus

Tests were conducted using the breadboard performance monitoring system with an operating microwave tube (2J52) and a 40-kW pulsed magnetron. **Figure 4.3** shows a simplified functional diagram of the overall experimental setup. During the tests, the magnetron RF output was transmitted via a waveguide system through several sampling modules to a dummy load. The directional couplers sent samples of the RF energy into sensors to measure the incident and reflected power, power output, and operating frequency. An adjustable mismatch, installed close to the dummy load, was used to vary the output voltage standing wave ratio (VSWR). Crystal detectors were used in all of the RF sensor circuits. During the experiment, the HP-1000 computer provided experimental control, data analysis, and data storage that would normally be provided by a microprocessor mounted in the crate. In addition, the computer system provided valuable capabilities to display data and to generate multicolor plots of experimental results.

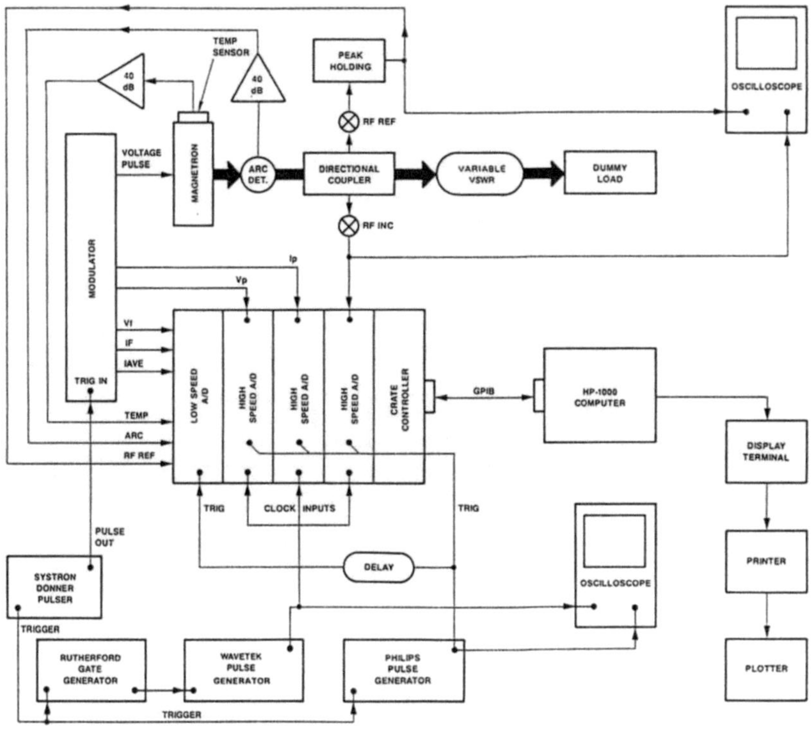

Figure 4.3.
Block diagram of experimental setup (after Watson and Troy [2]).

4.2.2 Experimental procedure

First, the magnetron was brought up to full power and allowed to stabilize. Power was then applied to all test components, and the experiment was initiated by activating the test control software in the computer. The test control program prompted the test operator to define sensors and to provide control of the experimental process. The computer program initiated the data acquisition process, and it then monitored and analyzed the RF sensor output for indications of faults. Normally, the tube would not fault without help from the operator. The test operator could perform the series of experiments by adjusting the environment or the terminal values to stress the magnetron tube. The magnetron was then stressed by turning up the pulse voltage or by turning off the magnetron cooling. Stress was also introduced by inducing high reflected energy in the output waveguide. When a particular type of tube stress was applied and a fault occurred, the computer would sense the fault, stop the sensing process, and store the fault data.

The data would then be transferred to the computer where it could be stored for later analysis.

The operator could then view selected pulse samples to verify the nature of the fault. By calling up the plotting program, the test operator could generate printed output and multicolor pen plots of the fault data.

4.3. Experimental results

An extensive set of measurements were conducted to generate fault data for analysis. These tests provided high-resolution, multisensor data from the pulsed

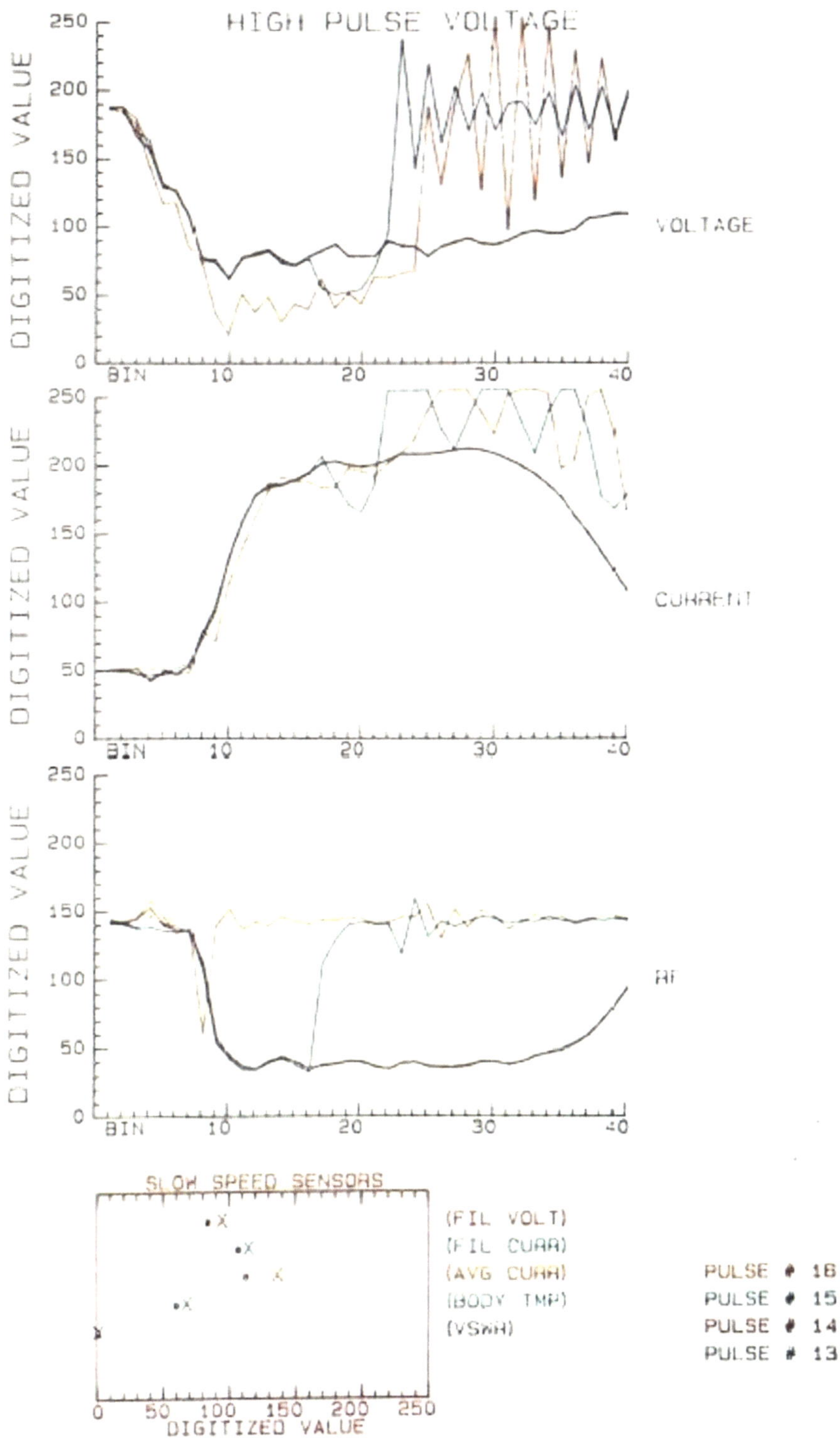

Figure 4.4.
Fault in magnetron caused by increased RF pulse voltage (after Watson and Troy [2]).

magnetron under stress. The analyses of these data were used as a basis for evaluating the effectiveness of the overall performance monitor concept. Two separate sets of measurements were conducted for each tube stress mode.

4.3.1 High pulse voltage

The high-pulse voltage tests were performed by operating the magnetron near full power and then slowly increasing the pulse voltage level until a fault occurred. Results from a typical test are presented in **Figure 4.4**. The computer plotter provided color plots of the pulse voltage, pulse current, and RF power output for several pulses before, during, and after the fault. All high-speed plots include at least one pulse before the fault and at least one or two pulses after each fault. The horizontal scale on these plots provides 40 resolution bins, corresponding to a time period of 2 ms, while the vertical scale shows the 256 amplitude levels of the 8-bit resolution of A/D converters. A summary plot for low-speed data is provided at the bottom of each figure.

The low-speed data were recorded during, or slightly after, every system pulse. The low-speed sensor display showed relative levels of the filament voltage, filament current, average tube current, body temperature, and VSWR. The display was designed as a horizontal bar chart, where a horizontal line is drawn to indicate the normal range of values and a tick mark to indicate the mean value. In addition, each low-speed sensor value at the time of the fault is marked with an "X." The VSWR values shown are peak values that were smoothed and filtered with a capacitive filter with a discharge time constant approximately four times the pulse length. In certain cases, the low-speed data would directly indicate the cause of the fault. In other cases, the low-speed sensors showed some of the effects from the fault. In many cases, results of low-speed sensors did not help identify the cause of the fault.

Returning to **Figure 4.4**, the results show that the initial fault occurred on pulse 15 (the blue curve). It started near the middle of the pulse with an increase in magnetron current and a resulting drop in pulse voltage. The drop in pulse voltage was due to increased modulator loading as a result of the tube arc current. Fluctuations of pulse voltage are a common characteristic of modulators operating under varying load conditions. This overvoltage is exhibited by the higher-than-normal-pulse voltage following the fault. The increased current could be caused by spurious emission from surface damage resulting from the arc. The severe drop-off of RF output during the middle of this pulse shows that the current flowing in the arc stopped the normal operation of the magnetron.

Low-speed sensors do not add much to the analysis of the test. The average magnetron current increased as expected during an arc. The changes in body temperature are minor and probably result from arc transients. Note that the low-speed sensor value labeled "VSWR" is only useful for indicating reflected power when the output power remains constant. During most experiments, the reflected RF energy would normally be expected to drop because of the drop in RF output. In addition, there would be fluctuations due to arc transients and spurious RF outputs, since the reflected power is measured during the pulse. Either of these effects can cause a significant variation in the reflected RF energy. In general, reflected energy is only considered as a valid indicator when it is the primary cause of the fault.

4.3.2 Low filament power

The low filament power tests were performed by bringing the magnetron up to full power, letting it stabilize, and then turning off the filament voltage. One of the problems with these tests was due to the difficulty in getting the magnetron to undergo stress. Since the magnetron was operating at moderate duty levels, there

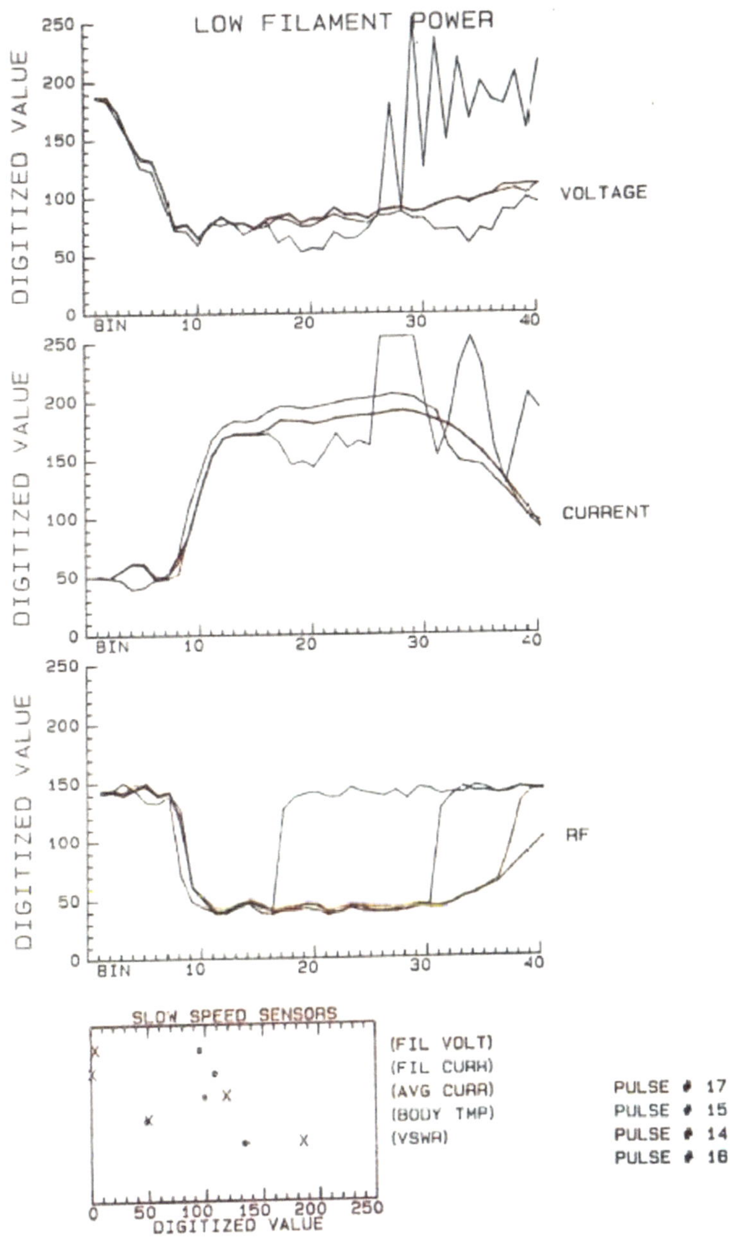

Figure 4.5.
Fault in magnetron caused by low filament power (after Watson and Troy [2]).

was a considerable amount of backheating due to RF interactions. It was therefore necessary to wait a long time after removing the filament power before any effects could be observed. The first indication of tube stress was a drop in magnetron current (**Figure 4.5**). This drop in current caused some overvoltage due to unloading of the modulator, which in turn induced an arc. An increase in pulse current on the following pulse was possibly a result of surface damage from the arc. The low-speed sensors clearly indicate the cause of the problem, since they show a major drop in filament voltage and current. Therefore, in this case, both high- and low-speed sensors are effective in pinpointing the cause of the problem.

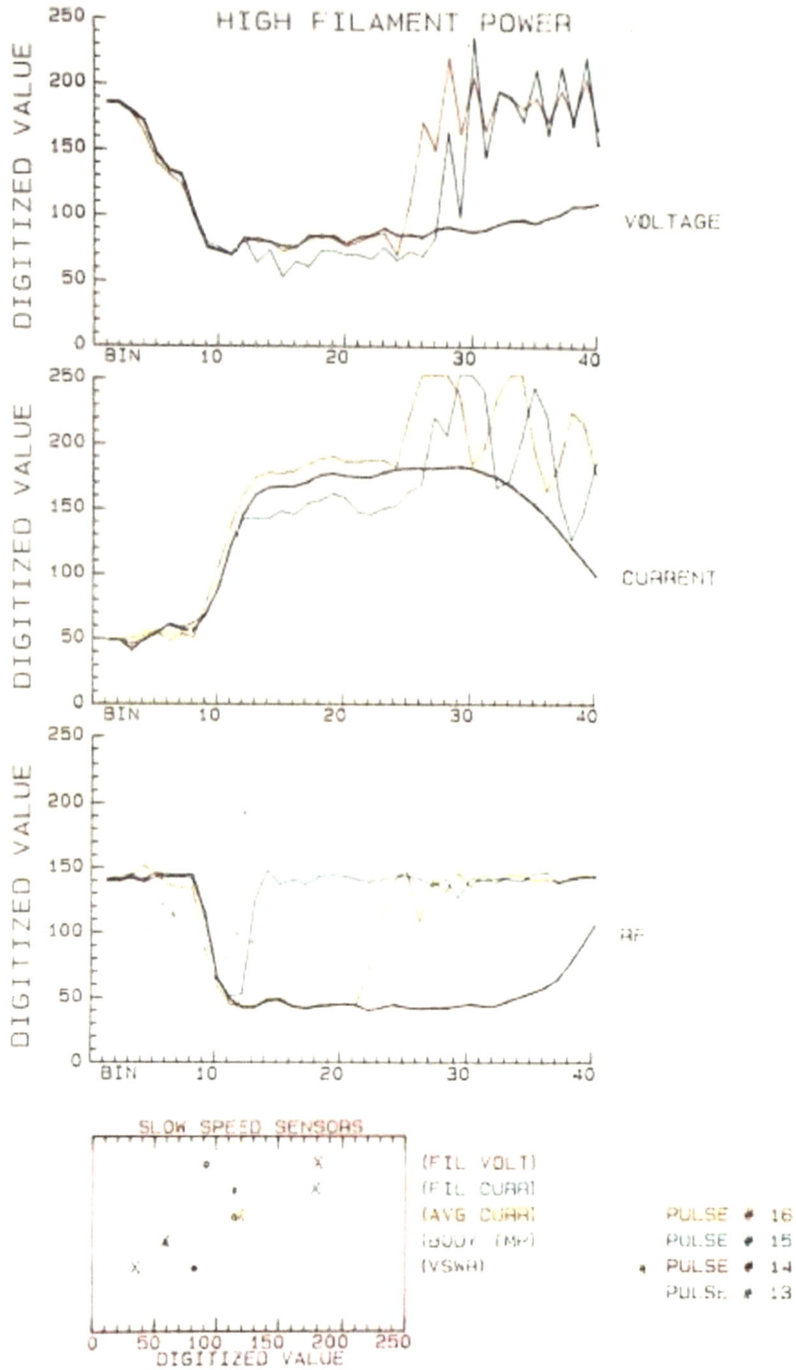

Figure 4.6.
Fault in magnetron caused by high filament power (after Watson and Troy [2]).

4.3.3 High filament power

These tests were performed similar to the previous tests, except that the filament voltage was turned up rather than down. **Figure 4.6** shows results very similar to those obtained when the filament power was turned off. The fault seems to start with a reduction in magnetron current, coupled with some overvoltage, and a

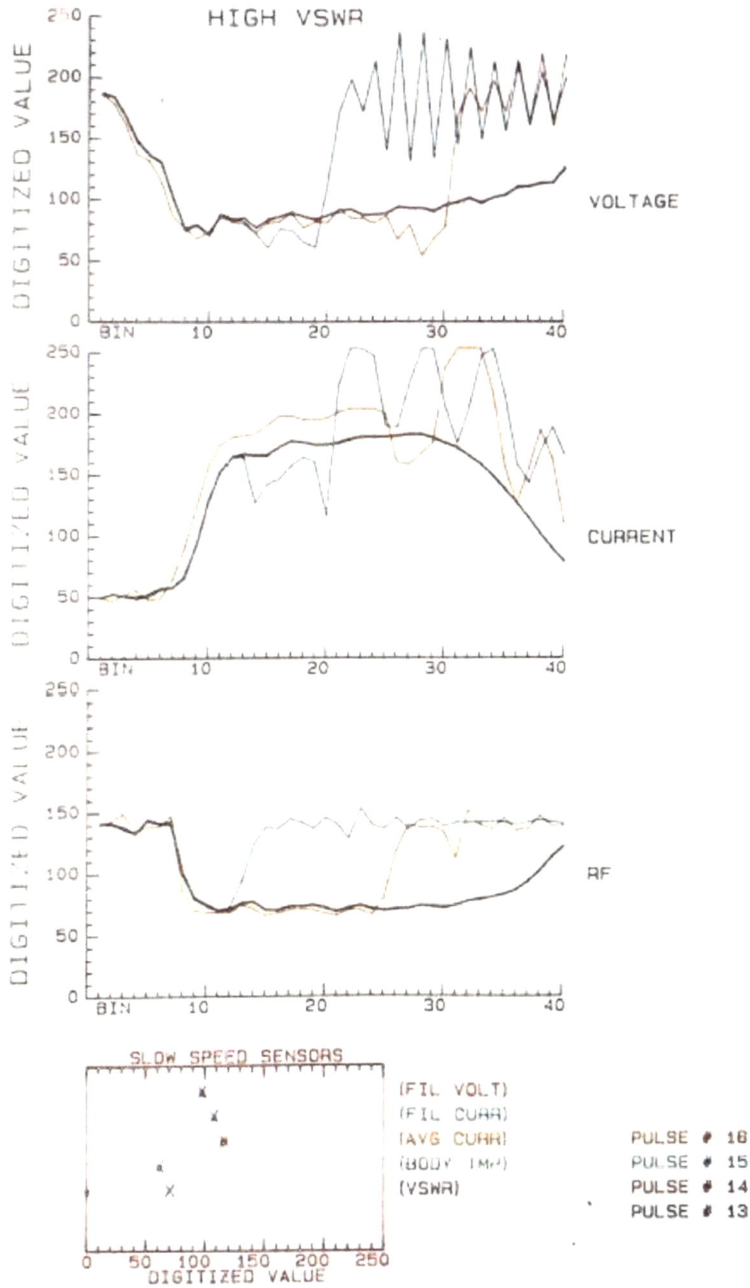

Figure 4.7.
Fault in magnetron caused by high VSWR (after Watson and Troy [2]).

subsequent arc. The basic failure mechanism seems to involve poisoning of the cathode by the evaporation of contaminating materials, as a result of the filament power increase. The low-speed sensors clearly indicate the cause of the problem is that the filament voltage and filament current are high.

4.3.4 High VSWR

The high-VSWR tests were performed by bringing the magnetron to full power and then adjusting the mismatch section in the output waveguide to introduce a large reflection. A high VSWR (approximately 4:1) was required to stress the magnetron sufficiently to cause it to arc. Results in **Figure 4.7** show that the initial fault starts with a reduction of magnetron current. This appears to be caused by high cathode back bombardment and damage resulting from the high fields generated

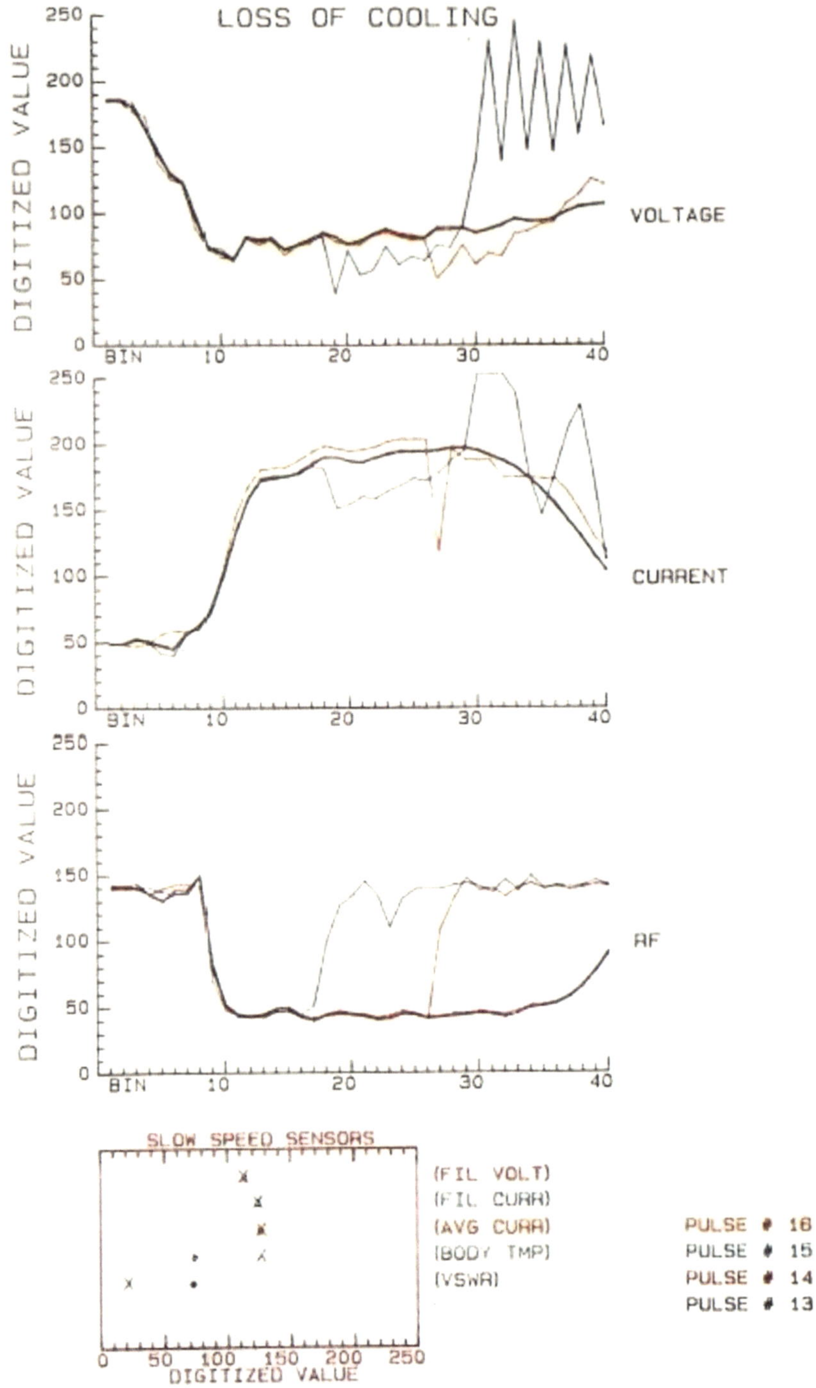

Figure 4.8.
Fault in magnetron caused by loss of cooling (after Watson and Troy [2]).

within the magnetron interaction area by the reflected energy. The low-speed sensor display pins down the actual cause, since high-VSWR values are clearly indicated.

4.3.5 Loss of cooling

The magnetron was cooled by an air blower. These tests were conducted by bringing the magnetron up to full power and then turning off the air cooling. It took a considerable amount of time, but the magnetron finally faulted due to overheating. **Figure 4.8** shows the results from these tests. Again, the fault is initially exhibited by a drop in emission, with the familiar sequence of over-voltaging and arcing. The low-speed sensor data can pinpoint the cause, since the body temperature reading is considerably above its expected value.

4.4. Summary and conclusions

Results obtained from the breadboard system clearly indicate that an effective microwave tube performance monitoring system could be fabricated using off-the-shelf components available at the time. Real-time data from a stressed (2J52) magnetron offered high-quality information previously unavailable for fault analysis. This kind of information, if returned with a failed tube, could be invaluable in determining the cause of failure. Furthermore, this kind of information can be used to identify tube and system deficiencies and system/tube incompatibilities.

The breadboard studies showed that high-speed sampling, without filtering, is adequate for almost all types of performance monitoring applications. Higher A/D converter speeds are becoming available, but higher sampling rates will not necessarily provide significantly greater analysis information. The time required to digitize data was not the principal limiting factor in applying these concepts. A major limitation resulted from the small amount of time available during the system interpulse period for data buffering and fault analysis. The breadboard system described here could only be used for radar tubes in the field operating at repetition rates of 200 pps or less. Increasing data acquisition speed could, in some cases, make the situation worse, by increasing the amount of data that must be transferred and analyzed during the small time interval available.

4.5. Some shortcomings and solutions

Several technical problems arose during the breadboard efforts. One of these involved sampling bin shifting when measuring repetitive pulses. This problem was eventually solved by using a triggered clock to cause the first sample-clock pulse to occur in synchronization with the system trigger. This type of synchronization was clearly necessary when performing high-resolution analysis of a sequence of pulses.

Another set of problems occurred because the microprocessor unit was operated in a high man-made noise environment. The solution to these problems was more difficult for sensors having very low signal levels, which made them more susceptible to interference from arcs and transients. The breadboard system layout was not ideal for minimizing this problem, since the components were spread out all over the test bench. Even so, it was possible to achieve reasonable success in reducing the effects of noise generated by fault transients and transmitted through space, and through the interconnecting power system. It was possible to solve these problems in the breadboard system by using power line filtering, shielding, and good ground connections. For example, the entire test bench was covered with a sheet of copper.

The ultimate solution to the interference problems was to mount all of the data acquisition modules in a single, shielded enclosure with a built-in power supply. The breadboard system was then successfully operated in a 40-kW pulse-power environment. When using a single-shielded enclosure, it was expected that satisfactory performance could be achieved in megawatt-level environments.

The interference problem was most difficult for the temperature sensor, which had a very low output, and required considerable amplification. The level of most of the other sensor inputs was greater than 20 or 30 mV; however, the problems for them were not too severe. The slowly varying data from the low-speed sensors were smoothed using low-pass filtering to eliminate spurious signals. Interference to the low-speed sensors was also effectively reduced by starting the A/D cycle after the end of the high-power pulse, thus effectively moving most of these signals out of the arcing time frame. The voltage levels used for triggering and synchronization were sufficiently high to make them reasonably immune from interference. Therefore, the use of other high-level logic systems would seem unwarranted for these kinds of applications.

4.6. Performance monitoring systems for traveling wave tube and crossed-field amplifiers

The monitoring system designs developed by Watson and Troy, and initially applied to magnetron tubes, were also adapted for performance monitoring of traveling wave tube and crossed-field amplifiers. Tests carried out on these devices gave similar successful results [3–5].

4.7. Conclusions

Based on these efforts and the results obtained, it seems clear that high-speed, microprocessor-based performance monitors using the technology of the time could be effectively used to improve microwave tube reliability. The performance monitor could be employed to localize and identify the nature and extent of problems that lead to eventual tube failures. Timely awareness of the problems that result from stressing microwave tubes beyond specified limits can prevent tube failures and significantly reduce annual operating tube costs for many systems currently used in the field.

References

[1] Watson WH, Troy JW. Reliability/Maintainability Monitor and Control for Microwave Tubes. NOSC Preliminary Report; 1980

[2] Watson WH, Troy JW. Computer-controlled performance monitor for microwave tubes. In: NOSC Technical Document 1070. San Diego, CA: Naval Ocean Systems Center; 1987

[3] Utilization of computer-aided real-time monitoring techniques for stress and reliability analysis. In: Proceedings of the Monterey Power Tube Conference, Naval Postgraduate School, Monterey, CA. Palo Alto, CA: Varian Corp.; 1984

[4] Rovnak JM. Traveling wave tube monitoring and control study program. NESC Contract No: N00123-79-C-0484, Final Report, Phase I. Fullerton, CA: Hughes Aircraft Co.; 1980

[5] Ashley GH. Reliability and maintainability monitor. NOSC Contract-No: 0058/NOSC Report No: CR-300. Waltham, MA: Microwave and Power Tube Division, Raytheon Co.; 1985

Chapter 5

Acoustic Emission Sensing for Microwave Tubes

5.1. Need for new nondestructive evaluation (NDE) methods

Current microprocessor-based systems, with as many as 11 sensors, are designed as described in the previous chapter, to monitor tube performance, provide tube protection, and record a comprehensive tube failure history. A major limitation with these systems is due to the small amount of time available during the tube's interpulse period, for data buffering and fault analysis. These monitoring systems work well if the microwave tube is operated at a rate of 200 pps or less. Microwave tubes are typically operated at rates up to 1000 pps with a 1-μs pulse duration. Increasing the data acquisition speed will, in this case, make the situation worse, since it increases the amount of data that must be transferred and analyzed during the small time interval available. Consequently, although the microprocessor-based monitoring systems have merits, they also have limitations. So a need exists for a different NDE solution that can overcome the limitations created by fast pulse repetition rates.

Research over the past several years has established the effectiveness of acoustic emission (AE)-based sensing methodologies for machine condition analysis and process monitoring. Acoustic emission has been proposed and evaluated for various sensing tasks, as described in the review of Chapter 1. These examples led us to try AE as an advanced NDE method for in situ performance monitoring of high-power microwave tubes. The question was whether the microwave radar tubes operating under normal conditions could generate AE signals. If they could not generate AE signals, then we could not use the technique. However, if they generate AE activity, we would need to understand the mechanism involved.

A literature survey led us to many research papers on topics indirectly related to the subject of the current investigation. We learned that acoustic emission was recently used to detect arcs on sputtering targets. Conventional arc detection techniques involve detecting voltage or current spikes occurring in power supply lines when arcs are initiated. These techniques are not immune to unwanted electrical noise signals and may generate false counts. Therefore, the implementation of nonelectric arc detection and counting techniques may provide some advantages by improving noise immunity. An arc detection and counting technique, which monitors the AE signals generated by the arcing event, was used successfully.

The fundamental principle of this method is based on the phenomenon of acoustic pulse generation when an arc-generated shock wave strikes a solid. The characteristics of AE signals generated by single-pulse corona discharges were measured and used as a diagnostic tool to estimate insulation deterioration in oil-immersed pole transformers. Frequency-spectrum analysis of AE signals generated by partial discharges was used to develop a technique to distinguish them from AE signals generated by corona discharges. For recorded waveforms of AE signals, Theobald et al. [1] pointed out the difference between the AE event energy (AE_{energy}) and the AE energy descriptor (E_{AE}). The AE event energy (AE_{energy}) is given by:

$$AE_{energy} \propto \int_{0}^{T} V(t)^2 dt \tag{1}$$

Generally, AE stress waves disperse throughout a medium until their interaction with an interface or boundary leads to the production of reverberating fields. Although this energy will be mostly absorbed by the medium, some of the energy can be detected by using a high-frequency piezoelectric ceramic transducer. Assuming that all mechanisms of energy loss in the structure and measurement system are constant, the measured electrical signal energy from the transducer could then be said to be proportional to the AE event energy (AE_{energy}). Here, T is the time length of the transient signal produced and V(t) is the transient signal voltage. The measured electrical signal energy is often referred to as the AE energy descriptor (E_{AE}) and is written as:

$$E_{AE} \propto \left(\frac{1}{R}\right)\int_{0}^{T} V(t)^2 dt \tag{2}$$

where E_{AE} refers to the AE energy measured in the transducer and R is the total impedance of the measuring circuit. The power of the AE signal of a detected event is thus proportional to the power of the source event. The advantage of energy measurement over ring down counting is that energy measurements can be directly related to important physical parameters without having to model the AE signal. Energy measurements also improve the AE measurement when emission signal amplitudes are low.

5.2. Application of AE sensing methods to microwave tubes

Based on these previously demonstrated AE sensing capabilities, we decided to try applying AE techniques as a new method for performance monitoring of high-power microwave tubes. A brief description of the equipment assembled together for the project is as follows.

5.2.1 Acoustic emission transducers

Four transducers were used in this study. They were types R50, R80, and S9208 from Physical Acoustics Corporation and B1025 from Digital Wave Corporation (**Figure 5.1**). The truncated conical aluminum piece at the right end was fabricated in a machine shop to function as a match between the curved surface of the anode of the magnetron and the flat front surface of the S9208 sensor. The R50 sensor has peaks at 100, 300,

Figure 5.1.
Acoustic emission transducers for AE signals from high-power microwave tubes. The rightmost object is aluminum matching buffer piece.

and 500 kHz, based on the data provided by the manufacturer. It is a general-purpose transducer useful for monitoring AE activity in the range of 100–700 kHz. The R80 sensor shows peak responses at frequencies of 130, 300, and 390 kHz according to data from the manufacturer. It is a general-purpose transducer useful for monitoring AE activity in the range of 200–1000 kHz. The S9208 sensor is a wideband displacement sensor recommended by the manufacturer for usage over the range of 20–1000 kHz. It is highly damped, using a tungsten powder backing. The fourth sensor used was a B1025 broadband sensor with a flat response over the frequency range of 1 Hz–1.5 MHz. The characteristics of the four transducers were independently evaluated during this work and were found to be in agreement to those supplied by their manufacturers.

5.2.2 Acoustic emission preamplifier

The 2/4/6 preamplifier from Physical Acoustics Corporation, used in this project, has three gain settings of 20, 40, and 60 dB. It has a high-input impedance, greater than 20 MΩ.

It can produce large output signals up to 20 Vpp into 50 Ω. The useful frequency range of this preamplifier as tested in this work is 100 kHz–1 MHz.

5.2.3 Acoustic emission postamplifier

A model AE1A from Physical Acoustics Corporation was used as the postamplifier. Its total gain, achievable in 3-db steps, is 41 dB. Its input impedance, at the terminal "AE INPUT," and output impedance, at the terminal "AE OUT," are both 50 Ω. Its useful frequency range, as tested in this project, is 100 kHz–1 MHz. The preamplifier receives about 21 V DC power from the postamplifier, on the same coaxial cable that carries the AE signal from the preamplifier to the postamplifier.

5.2.4 The oscilloscope

To view the AE activity of a microwave radar tube in operation, we decided to use a digital oscilloscope. The oscilloscope used to monitor AE activity was a LeCroy Model 9354AM. One important feature of this oscilloscope was its four channels, each having a bandwidth of 500 MHz. Each of the oscilloscope's four identical input channels was equipped with a 500-MSPS, 8-bit analog-to-digital converter. Acquisition memories of 2 million points simplified transient event capture by providing long-waveform records. The capture of waveform was possible even when the trigger timing or signal speed was uncertain. The data from the screen could be dumped and stored on a 3.5-inch, double-sided, high-density disk. This completes the description of the equipment.

5.3. Response characteristics of transducers

The frequency responses of all four transducers were evaluated in the laboratory although data on the responses had also been supplied by their manufactures. These manufacturer-provided responses are shown in **Figures 5.2A–5.5A** for the R50, R80, S9208, and B1025 transducers, respectively. The calibration of the AE sensors was standardized by using a capacitive transducer to capture the initial arrival of a displacement signal, produced by the fracture of a glass capillary on a metal surface. This procedure allowed the development of quantitative comparisons of AE experiments with theories of acoustic wave propagation. AE activity may be of a discrete or continuous nature. Attempts were made to generate standard pulses, which for discrete ones, corresponded to the delta function, while in the case of

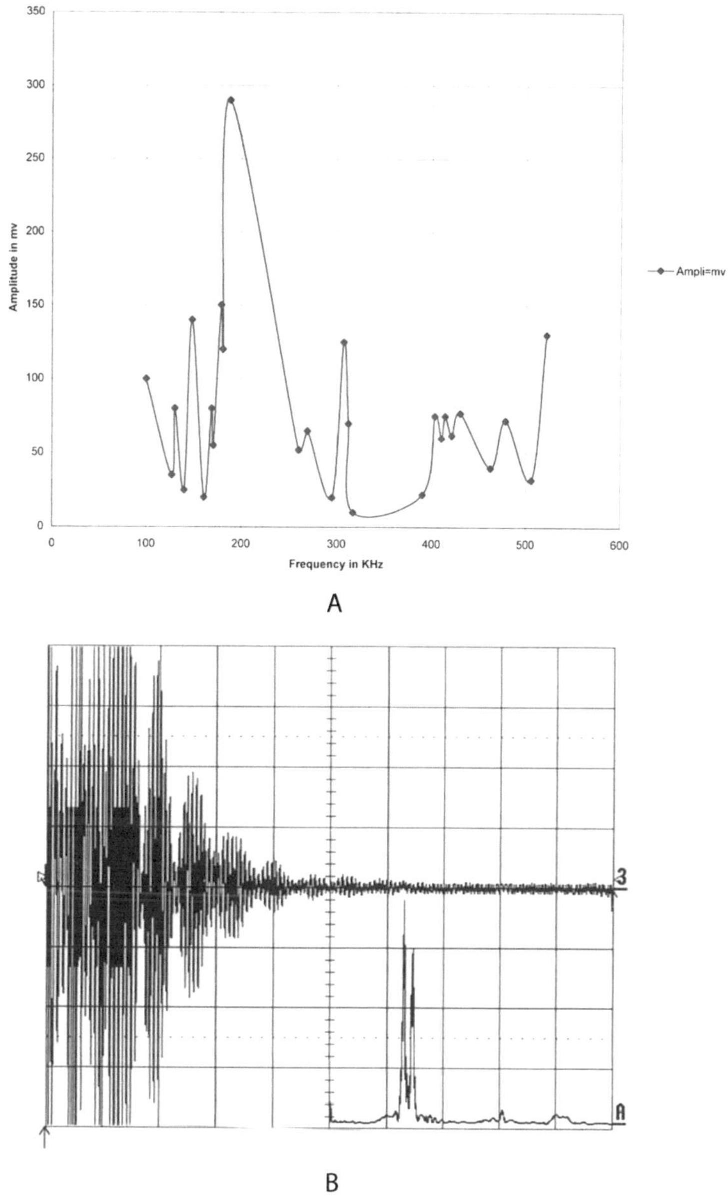

Figure 5.2.
(A) Amplitude versus frequency for transducer R50. (B) Response of R50 to impulse loading and spectrum (Ch 3: AE Signal; Time Scale =0.1 ms/div; Y-axis = 200 mV/div; Ch A: Magnitude of FFT; X-axis = 0.1 MHz/div; Y-axis = 17 mV/div).

continuous emission, to white noise. The standard signal should be characterized by parameters, measured in time and frequency domains, using sensors located as closely to the AE signal source as possible. The breaking of a mechanical pencil lead simulated an AE source of discrete nature. This procedure was performed on a circular aluminum plate one-half-inch thick and 6 in in diameter. The lead was broken on the plate at a distance of ½ in from the AE transducer, which was coupled with silicone grease to the same surface of the plate. **Figures 5.2B–5.5B** show the responses of sensors R50, R80, S9208, and B1025, respectively, to the laboratory AE simulation testing. In all four of these figures, Channel 3 (horizontal scale, 0.1 ms) shows the received AE signal, while trace A (horizontal scale 100 kHz per division) shows the magnitude fast Fourier transform (FFT) of the AE signal.

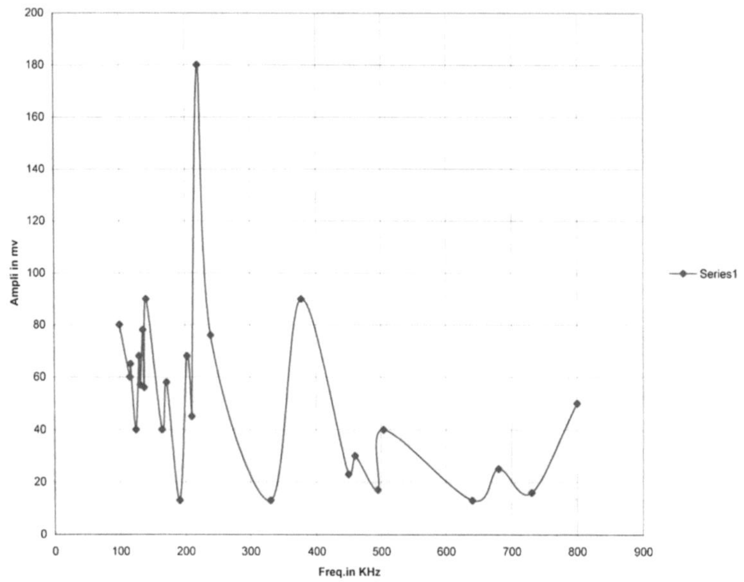

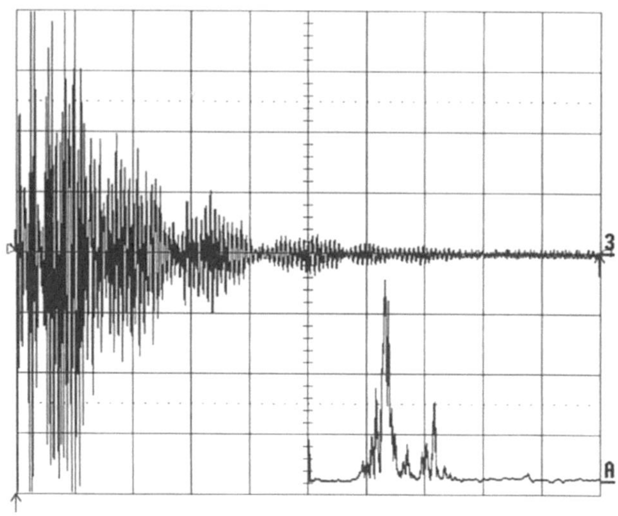

Figure 5.3.
(A) Amplitude versus frequency for transducer R80. (B) Response of R80 to impulse loading and spectrum (Ch 3: AE Signal; X-axis =0.1 ms/div; Y-axis = 0.50 V/div; Ch A: Magnitude of FFT; X-axis = 0.1 MHz/div; Y-axis = 23.5 mV/div).

Comparing **Figure 5.2A** with **Figure 5.2B**, both for the narrow-band sensor R50, **Figure 5.2A** shows that there are five peaks between 100 and 200 kHz, while on the same frequency range, two peaks can be easily identified in **Figure 5.2B**.

Next, comparing **Figure 5.3A** with **Figure 5.3B**, both for the narrow-band sensor R80, **Figure 5.3A** shows two peaks between 200 and 300 kHz, while over the same frequency range, two peaks can also be identified in **Figure 5.3B**. Comparing **Figure 5.4A** with **Figure 5.4B**, both for the broadband, highly damped transducer S9208, the peaks in **Figure 5.4A** are not high, although we see them relatively more pronounced in the region between 90 and 200 kHz, while **Figure 5.4B**, on the contrary, shows much higher peaks on the same frequency band relative to peaks at other frequencies. For this sensor, smaller peaks can be found up to 500 kHz.

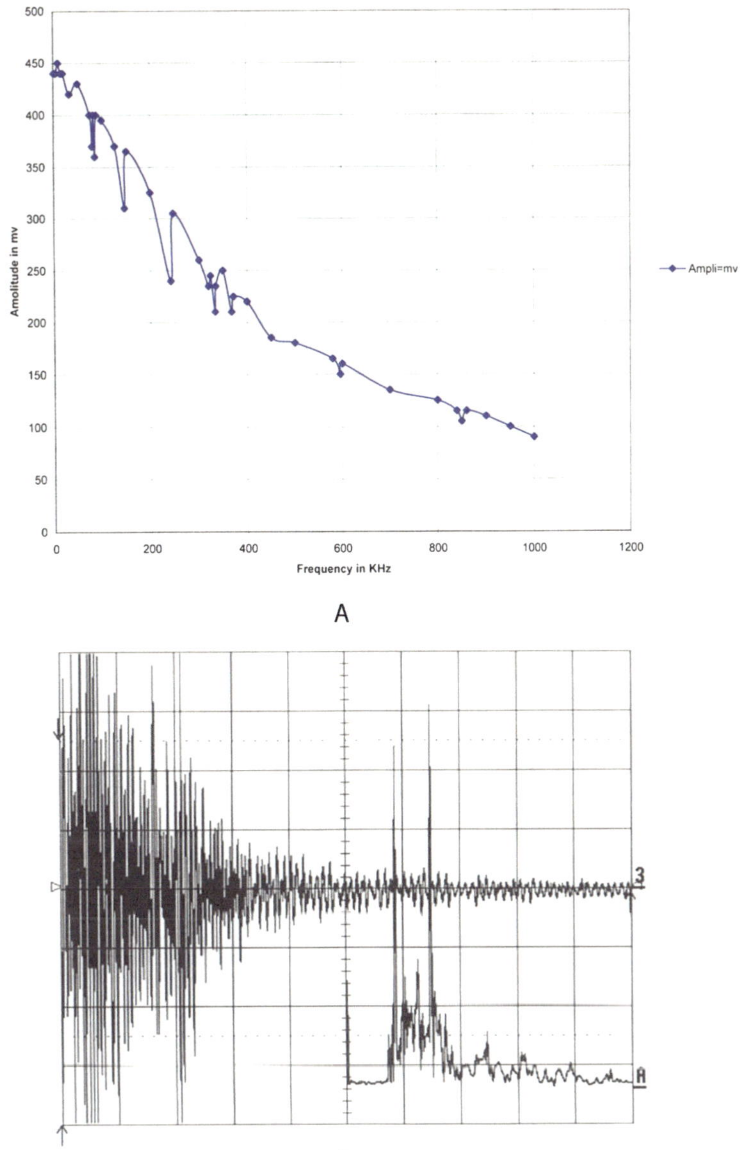

Figure 5.4.
(A) Amplitude versus frequency for transducer S9208. (B) Response of S9208 to impulse loading and spectrum (Ch 3: AE Signal; Time Scale =0.1 ms/div; Y-axis = 0.50 V/div; Ch A: Magnitude of FFT; X-axis = 0.1 MHz/div; Y-axis = 20 mV/div).

Finally, comparing **Figure 5.5A** with **Figure 5.5B**, both for the broadband transducer B1025, **Figure 5.5A** shows a continuous decrease in amplitude, without peaks, with increasing excitation frequency, while **Figure 5.5B** shows high peaks over the band of 100–200 kHz.

5.4. Applications of AE sensor technology to high-power microwave tube magnetron

The high-power microwave tube used in this experiment was a type 2J55 magnetron. A magnetron is an oscillator that is small and lightweight, with reasonable operating voltages, good efficiency, rugged construction, and a long life.

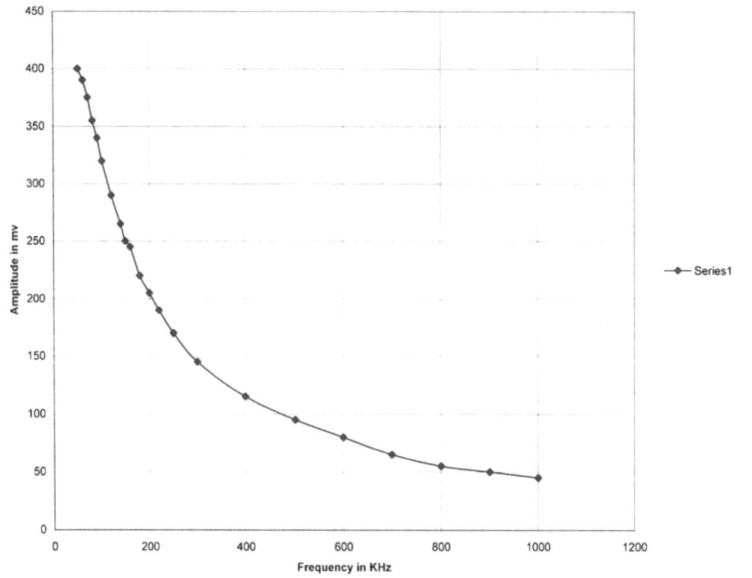

A

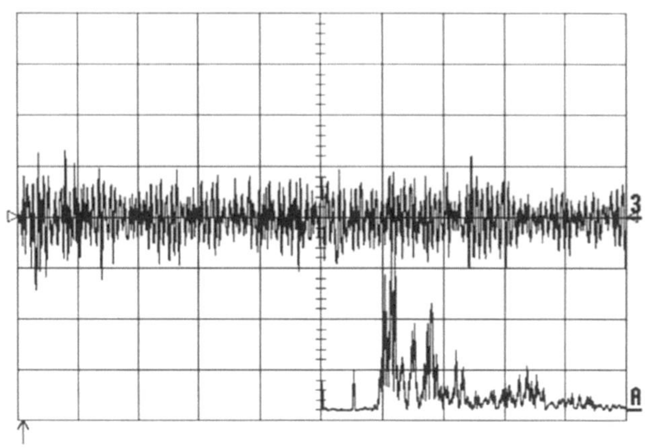

B

Figure 5.5.
(A) Amplitude versus frequency for transducer B1025. (B) Response of B1025 to impulse loading and spectrum (Ch 3: AE Signal; X-axis = 0.1 ms/div; Y-axis = 200 mV/div; Ch A: Magnitude of FFT; X-axis = 0.1 MHz/div; Y-axis = 10 mV/div).

The magnetron was the first practical high-power pulsed radar source used at microwave frequencies, and it is still rather widely used today. The magnetron converts energy, extracted from an electric current, to an RF field. It is a kind of diode [3], as shown in the simplified schematic of **Figure 5.6**.

The principal parts of a magnetron are the cathode, the interaction region, the cavities, and the output coupling. The magnetron essentially consists of a copper cylinder, into which resonant holes were machined. These holes connect with the interior of the cylinder via slots. A cathode is mounted concentrically in the cylinder and an axial magnetic field is applied to the whole structure. A typical field strength for this magnetic field, in the case of a 2J55 magnetron, is 1900–2100 Gauss. This magnetic field is normally provided by permanent magnets built into the magnetron, and is not changeable.

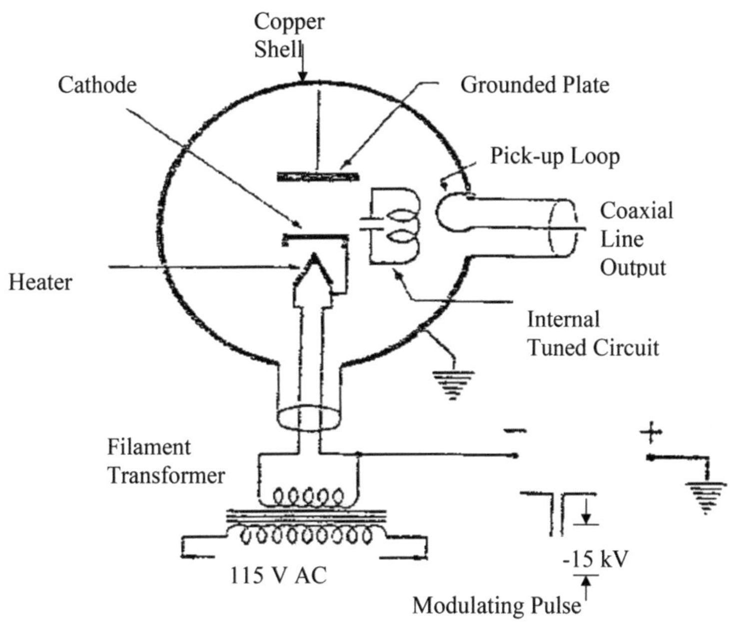

Figure 5.6.
Simple schematic of magnetron tube (after TM 11-1524 [3].

The outside of the magnetron is a grounded shell to which the plate of the magnetron is connected (**Figure 5.6**). The cathode of the tube is heated indirectly and has one side of its heating filament connected to it. The cathode heater filament voltage is 6.3 V. The filament is fed by a transformer whose secondary winding is insulated to withstand the peak modulator pulse voltage applied to the cathode. Applying a high negative voltage pulse, through a high-voltage concentric bushing, to the cathode of the magnetron leads to the emission of electrons from the cathode, as illustrated in **Figure 5.6**. The electric field E and magnetic field H are crossed in the interaction region; hence, the magnetron is called a crossed-field device. The electrons emitted from the cathode are initially accelerated toward the positively charged regions of the anode. In traveling toward the anode, the paths of the electrons are curved by the axial magnetic field and their trajectories take them toward the negatively charged regions of the anode. This leads to positive feedback and causes the magnetron to oscillate. In addition to their basic helical motions, the electrons are also velocity modulated and tend to form bunches as they pass the cavities. In an eight-cavity magnetron, these bunches form four spokes centered on the cathode and these rotate in synchronism with the oscillatory field. If the conditions are chosen correctly, the electrons will convert some of their kinetic energy into RF energy. This energy given up will increase the amplitude of the oscillations in the cavity.

When the magnetron is oscillating, the cavities are coupled together by oscillatory E and H fields, and therefore energy can be extracted from all cavities by a loop within a single cavity. The voltage/current relationship for a microwave magnetron is nonlinear, and hence, a magnetron is often represented by an equivalent circuit consisting of a biased diode with the magnetron's stray capacitance represented by a capacitor connected in parallel. An appreciable current is not drawn by a conventional magnetron until a critical voltage, approximately 90% of the operating voltage, is reached. This voltage is called the Hartree voltage [2]. However, low levels of RF output power may be generated by the tube, even below the Hartree voltage. Therefore, be careful to ensure that voltage is promptly removed after the desired pulse and not reapplied, even at low levels.

5.5. Magnetron operating characteristics

A photograph of a magnetron tube along with other microwave tubes is shown in **Figure 5.7A** and **B**. A magnetron unit is mounted on top of two large cabinets

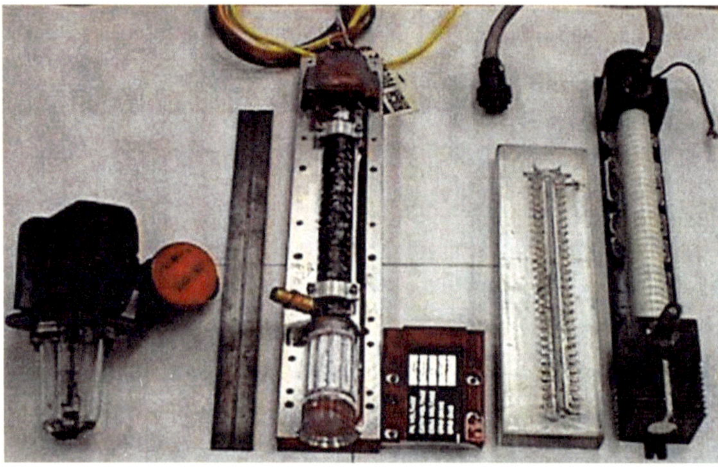

A

B

Figure 5.7.
(A) Magnetron, ruler, opened TWT, and TWT (from left to right). (B) Klystron amplifier electronic Tube from varian-type VA-889F.

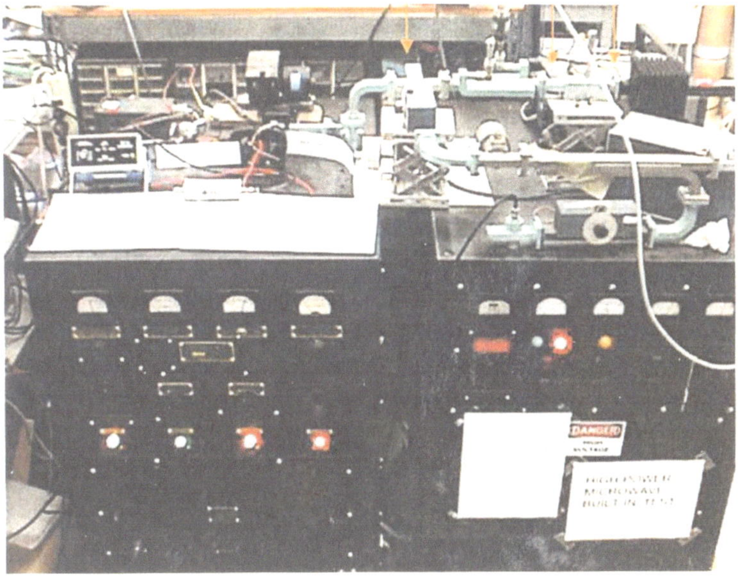

Figure 5.8.
Magnetron (center) with waveguide (right) on the top of power supply and pulser cabinets.

containing a power supply and modulator as shown in **Figure 5.8**. When a voltage pulse of the proper magnitude, typically −15 kV DC, is applied, the magnetron will oscillate at a frequency determined by the dimensions of the cavities in the magnetron, which form a resonant circuit. The RF energy generated by the magnetron used in this setup was at 9.6 GHz. For a 2J55 magnetron, the size of the cavities is around 2 mm. As mentioned earlier, one cavity is coupled inductively to a pick-up loop that is connected to a probe transition that feeds the RF energy into a waveguide. The peak electrical input power to the magnetron was 210 kW and the peak RF output power was 50 kW. The current during the 1-ms pulse is about 20 A. The duty cycle (width of pulse/period between pulses = 1 ms/0.01666 s) for 60 pps is 6×10^{-5}, or 0.006%. This is also called the duty factor (pulse length × pulse repetition rate = 1 ms × 60 pps). Combining these results gives a value for the average current (peak current × pulse width × number of pulses per second = 20 A × 1 ms × 60 pps, for example) drawn by the magnetron of about 1.2 mA.

One problem with magnetron oscillators is that the interaction space in a magnetron can support more than one possible mode of oscillation. In well-designed magnetrons, the various modes are well separated in both voltage and frequency, and the magnetron will provide stable operation in the desired mode. However, certain parameters, notably the characteristics of the modulator, such as the rate of rise of voltage (RRV), can strongly influence the mode of oscillation in magnetrons. Operation in other than the normal mode (called moding) is usually undesirable because oscillations are not at the normal operating frequency, and consequently, the power does not properly couple out of the tube. Double moding can be detected by the appearance of a smaller pulse, which varies in amplitude, underneath the main pulse. Double moding could also happen due to a change in loading, in magnetic field strength, or in applied voltage [3].

Another property of magnetrons, which is probably related to moding, is their tendency to produce occasional RF output pulses of reduced amplitude or to fail to produce any RF output pulse. In a well-designed tube, such missing pulses would typically occur less often than 1 in 10^4. As with any high-power tube, the possibility of internal tube arcing always exists. Particularly, in the case of magnetrons, internal arcing is one factor that must be given attention in the design of the modulator or pulser used to drive the tube.

In some magnetrons, the impact of electrons returning to the cathode generates so much heat that a magnetron can continue to operate normally if the filament voltage is removed. The output frequency varies about 0.05 MHz per degree change in magnetron temperature. Thus, a gradual frequency shift may be observed during the initial warm-up period. The magnetron is cooled during normal operation by an air blower and its temperature is monitored by a temperature sensor connected to a digital readout device. As mentioned before, the resistance of the magnetron is nonlinear. At low plate-voltage values, large changes of plate voltage result in minor changes in plate current. In the normal operating mode, however, a 10% increase in plate voltage may produce a 100% increase in plate current, depending upon the characteristics of the tube operated at a given value of the magnetic field strength.

5.6. Waveguides and VSWR

For minimum losses and high-power transmission at microwave frequencies, waveguide is used extensively. The magnetron in the current work was connected to a rectangular waveguide. A waveguide consists of a single metallic conductor in the shape of a rectangular box, or cylinder, inside of which electromagnetic waves are propagated. Waveguides are usually made of brass, copper, or aluminum in various standard sizes corresponding to the microwave frequencies used. Waveguide sections are usually coupled together by flanged assemblies that are bolted together and supported periodically on metal stands. The guided waves in waveguides have field configurations that are somewhat different from those of two-wire open lines. Waveguide transmission line is characterized by two important features. First, there is a minimum frequency below which a given waveguide will not transmit the wave. This is called the cut-off frequency, fc, and is directly related to the size of the waveguide used. Second, there is always a component of either E or H along the direction of propagation. The guided waves propagate along the waveguide with different field patterns called "modes." Modes are of two types, those that have an E component along the direction of propagation, called TM or E waves, and those with an H component along the direction of propagation, called TE or H waves. Rectangular waveguides have fewer losses than those of circular waveguides and are less prone to develop undesired modes. Hence, they are commonly used. The parameters associated with rectangular or circular waveguide are impedance Z and cut-off wavelength λc, and they are closely related to the dimensions of the waveguide, while attenuation losses depend on these factors as well as on the inner surface finish and metal of the waveguide walls. For an example, some figures for the TE10 wave mode in rectangular waveguide are as follows:

$$\text{ZTE (Ohms)} = 377\lambda g/\lambda \text{ and } \lambda c = 2a, \tag{3}$$

where λg is the guide wavelength, λ is the free space wavelength, and λc is the cut-off wavelength with "a" being the width of the waveguide [4].

The magnetron can be connected to a waveguide to inject microwaves into which they can propagate as sinusoidal waves to an RF load. If the load absorbs all of the waves, then there is no reflection. This occurs when the RF load and the waveguide have equal impedances. This condition is called a matched load. But if there is a load mismatch, then some of the incident power is not absorbed by the load and is

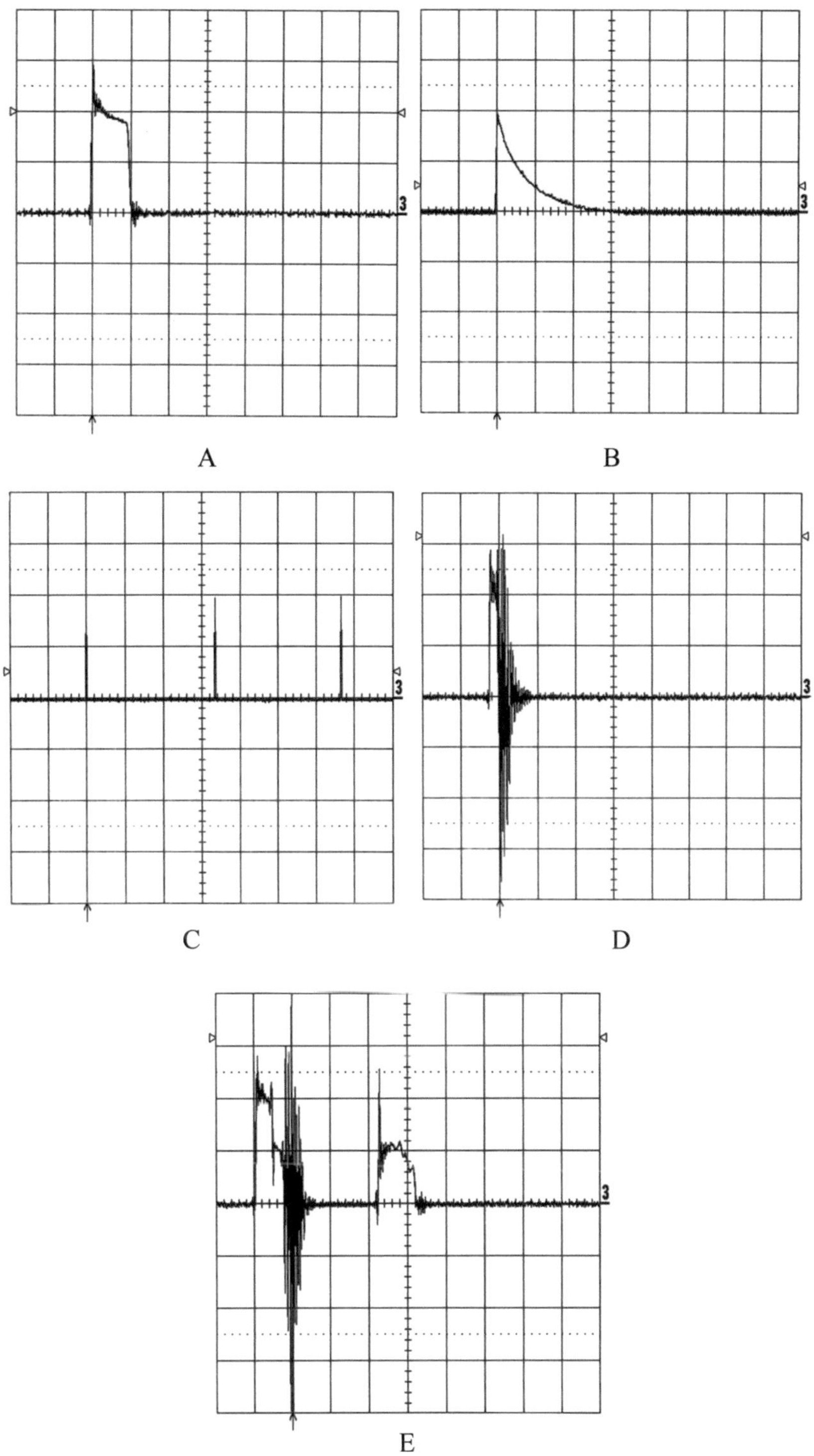

Figure 5.9.
(A) Good RF pulse matched coupling-50 Ω input impedance to the oscilloscope. (B) Distorted RF pulse unmatched input impedance of 1 MΩ to the oscilloscope. Note: Both pulses from the square law diode detector from the location on the waveguide of the magnetron. (C) Good RF pulses at 60 pulses per second and separated by 16.67 ms. Time scale = 5 ms/div, amplitude on Y-axis = 5 V/div. (D) Faulty RF pulse from magnetron with high frequency content. Pulse width = 1 μs X-axis = 1 μs/div Y-axis = 200 mV/div. (E) Two faulty RF pulses from magnetron due to irregular triggering. Time scale = 1 μs/div amplitude on Y-axis = 200 mV/div.

reflected back. The reflected wave then combines with the transmitted wave, resulting in a new wave pattern for which the amplitude varies with the position along the waveguide. These amplitude variations stand still in the waveguide and, hence, are called standing waves. The ratio of the maximum to the minimum amplitude is called the voltage standing wave ratio or VSWR, and is a measure of the amount of RF energy reflected back into the waveguide toward the magnetron. Knowing the VSWR at the load is important because if the VSWR is too high, an appreciable amount of energy will be reflected back to the magnetron, which could cause severe damage. The VSWR can be written as:

$$VSWR = V_{max}/V_{min} = (Vi + Vr)/(Vi - Vr), \tag{4}$$

where Vi and Vr are the voltages due to the incident and reflective wave, respectively. The VSWR can be measured with a network analyzer or with the slotted line method using a probe and VSWR meter [5].

5.7. RF pulses from a magnetron

Before coupling any AE equipment to the magnetron system, we decided to operate the magnetron and observe the RF pulses generated by it using the oscilloscope. **Figure 5.9A** shows a single RF pulse of 1-ms width with a 50-Ω terminating resistor at the input channel of the oscilloscope. **Figure 5.9B** shows the same pulse without the termination.

Figure 5.9C shows a time interval of 16.66 ms between successive RF pulses for a pulse repetition rate of 60 pps. Even during normal operation of the magnetron, faulty RF pulses could occur. **Figure 5.9D** shows a single bad RF pulse with a high-frequency content. **Figure 5.9E** shows two bad RF pulses with a separation less than 16.67 ms due to irregular triggering. These RF pulses were from the square law crystal diode detector. The input to the square law diode was used to detect the RF output from the waveguide of the magnetron. The height of the pulse could be adjusted using an attenuator inserted into the waveguide. The frequency could be measured using a frequency meter coupled to the waveguide. During this procedure, the whole test assembly was kept cool by two running fans, one for the magnetron and the other for the load at the other end of the waveguide. With these initial tests providing an understanding of the functioning of the magnetron, the next step was to couple AE equipment to the magnetron unit and observe and record AE signals generated during normal and abnormal operation of the magnetron, as described in the next chapter.

References

[1] Theobald P, Rokowski RT, Yan T, Jarvis D, Dowson S, Jones BE. Reference source for calibration of acoustic emission measurement. In: IEEE Instrumentation and Measurement, Technology Conference, Budapest, Hungary, May 21-23, 2001. pp. 412-416

[2] Ewell GW. Radar Transmitters. New York: McGraw-Hill Book Company; 1981. p. 23

[3] Radio Sets SCR-584-A and SCR-584-B Service Manual, Theory, Troubleshooting and Repair, War Department Technical Manual, TM-11-1524, July 1946, p.15

[4] Connor FR. Wave Transmission. London, UK: Edward Arnold; 1972. p. 7

[5] Love W. Magnetrons. In: Koryu Ishii T, editor. Handbook of Microwave Technology. Vol. 2. San Diego, CA: Academic Press; 1995. p. 36

Chapter 6

Acoustic Emissions from Magnetrons

6.1. Experimental setup

A simplified schematic of the experimental equipment is shown in **Figure 6.1**. **Figure 6.2** is a photo showing the 2J56 magnetron and the acoustic emissions (AE) equipment arranged on the top of two large cabinets housing a high-voltage DC power supply and pulse modulator. The magnetron is painted black, and is hence difficult to see it in the figure. The red-colored cable in the figure runs from the sensor mounted on the cylindrical anode of the magnetron to the input of the preamplifier located in the front. On the right-hand side is the waveguide (greenish-gray color) carrying RF energy to the load. On the left-hand side is the postamplifier. In Figure 5.8 of Chapter 5, the complete waveguide is seen with the square law crystal diode RF detector located adjacent to the RF load at the right end. The voltage pulses from this detector, representing the RF pulses, are sent to Channel 4 of the oscilloscope (**Figure 6.3**), the AE signal is displayed on Channel 3, the cathode current pulse is fed to Channel 2, and the cathode voltage from a high-voltage probe (Tektronix-Model 015-049) goes to Channel 1. In **Figure 6.3**, an Interstate Electronics Model P25 pulse generator is located under the oscilloscope. Its output (typically 200–1000 pps) goes to the external trigger on the oscilloscope and the pulse modulator trigger (left-hand cabinet). The pulser's pulse rate (number of trigger pulses per second), width, and amplitude can be adjusted while being monitored on the oscilloscope, for example on Channel 2. The oscilloscope can process the four traces A, B, C, and D in several ways, including zooming in the time scale. In this case, the top trace C on the oscilloscope screen is the zoom of an individual RF pulse from Channel 4, the middle trace is the AE signal on Channel 3 as described before, and the bottom trace B is the zoom of a single cathode current pulse from Channel 2. The trace A is used to obtain a magnitude fast Fourier transform (FFT) of the AE signal. The scope's capacity to test a waveform, in this case a single current pulse on trace B, against a defined mask (trace D), made it possible to capture on the screen abnormal current pulses and run mathematical operations such as FFTs of AE signals generated during abnormal functioning of the magnetron. As discussed in Chapter 4, the magnetron was purposely stressed by shifting the parameters from their normal values at which it was designed to operate. It may then produce abnormal cathode current pulses leading to abnormal RF pulses. The magnitude FFT of the AE activity picked up during abnormal functioning of the tube was computed by the oscilloscope and all traces from its screen saved on a disk. These traces were then compared with those obtained during stable functioning of the tube. This testing procedure was expected to show us whether the acoustic emissions from the magnetron could be used as a noninvasive method of performance monitoring of the normal and abnormal functioning of the tube.

6.2. Experimental procedure

The normal operating parameters of the magnetron (2J56) used in this work and deviations made in them to stress the tube are described next. The normal filament voltage of the oscillator tube is 6.3 V AC, obtained through a step-down transformer whose primary normally operates at 115 V AC. By decreasing the voltage of the primary from 115 to 60 V, the cathode emission could be decreased to produce abnormal current pulses. This was the first deviation from the normal parameters. The normal operating pulse rate (PR) of the pulses from the high-voltage modulator was 1000 pulses per second (pps). It could be increased up to 1500 pps or decreased to 100 pps. This was the second deviation from the normal parameters. The pulse width was maintained constant at 1 μs.

In this work, the pulse rate was varied within the range of 100–1000 pps. With 1000 pps, the average power output was around 50 W. With 100 pps, it decreases to 5 W.

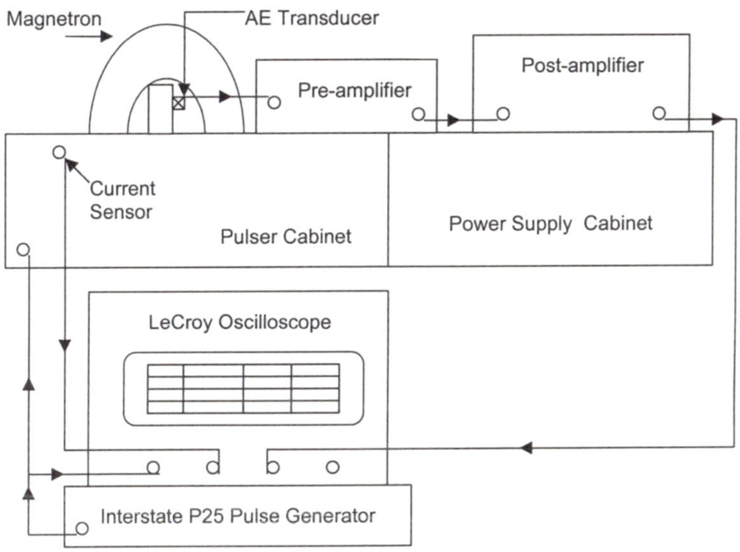

Figure 6.1.
Schematic of acoustic emission experiment with magnetron (after Joshi et al. [?]).

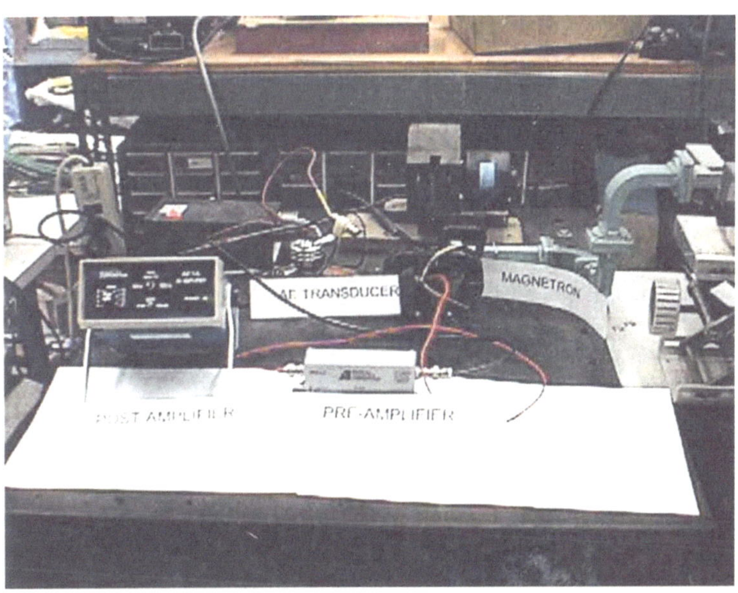

Figure 6.2.
Photo of setup of AE equipment next to magnetron.

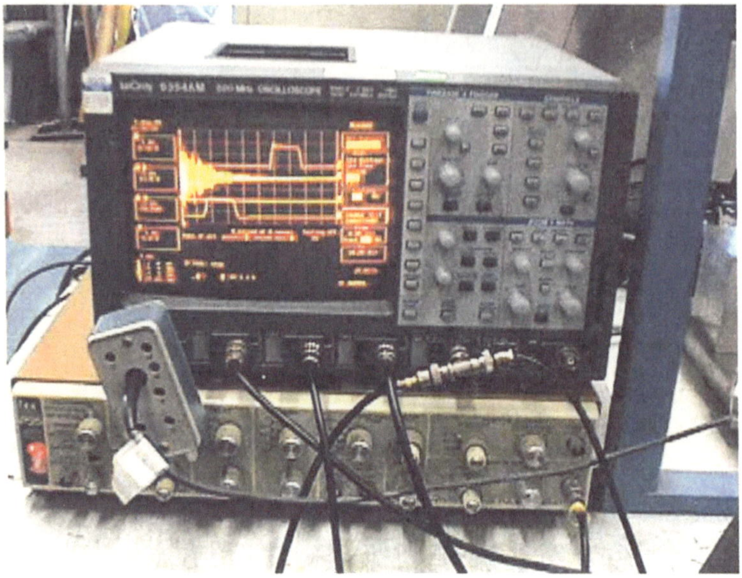

Figure 6.3.
AE signal from magnetron on LeCroy digital oscilloscope.

The ambient room temperature was about 65°F. At 100 pps, the temperature of the tube rose to 95°F. To keep it from rising beyond that level, air was blown on the tube with a fan, with a second fan keeping the waveguide section and load cool. The third way to stress the tube was to shut off the first fan and let the temperature of the magnetron rise. The temperature rise was rather rapid when using a pulse rate of 1000 pps.

The RF energy produced by the tube was delivered to a load at the other end of the waveguide. By using a variable impedance load attached to the waveguide, a part of the energy could be reflected back to the magnetron, creating an abnormal operating situation. Changing the ratio of the reverse power to the forward power was the fourth way to stress the tube. The forward and reflected power levels were measured with a Hewlett-Packard Model 432A power meter. Two 10-dB attenuators in series were inserted when connecting the power meter sensor to the waveguide directional coupler, when the power output was around 50 W, with only one attenuator used when the power output was around 5 W.

The normal pulse voltage (PV) from the modulator was 15 kV. It could be increased to 21 kV. This was the fifth way to create an abnormal operating parameter. However, this deviation could produce arcing among components inside the high power supply cabinets. For this reason, it was used sparingly in this effort. In general, when the tube was stressed, it produced abnormal current pulses, with resulting abnormal RF pulses. The abnormal current pulses were captured using the masking feature of the digital oscilloscope. The data acquisition was then stopped and a magnitude FFT was computed for the AE signal captured from the magnetron. The frequency spectrum of the AE activity was expected to be less than 1 MHz. Consequently, the sampling rate was set to 250 MS/s. The displays on the oscilloscope screen were saved on a computer disk by using its screen dump function. This procedure was carried out for both normal and various abnormal functioning of the magnetron tube. The AE activity was monitored with the four different types of transducers, described in Chapter 5, in turn. Out of the 100 acoustic emission data displays reported in SSC Pacific Technical Report 1810 [1], only a selected few are reported here to prove the usefulness of AE techniques.

6.3. Experimental results

The data in **Figure 6.4A** and **B** were obtained by using the S9208 AE transducer. **Figure 6.4A** and **B** shows the masking feature of the LeCroy digital oscilloscope and its ability to catch any abnormal operation of the magnetron when it was stressed by changing one of its normal operating parameters. The oscilloscope was triggered to record its traces any time a trace crossed outside of the boundaries of its mask. In both figures, the pulse rate was adjusted to 100 pps. The oscillator filament primary voltage was set at the normal value of 115 V so that the secondary received 6.3 V. In both figures, oscilloscope Channel 3 was connected to the output of the acoustic emission postamplifier. Channel B was connected to the current sensor of the magnetron, and was set to show the zoom trace of the current pulse. Channel D shows a zoom trace of the mask used for the current pulses of Channel B. Channel A shows the magnitude FFT of the acoustic emission signal of Channel 3. In **Figure 6.4A**, the pulse voltage was set at the normal value of 15 kV, while in **Figure 6.4B**, it was gradually increased to 19.5 kV until the abnormal current pulse, shown in the figure, occurred. This abnormal pulse was caught by the mask function, which in turn triggered a screen dump onto the floppy disk in the drive in the oscilloscope.

In comparing the two AE signals of **Figure 6.4A** and **B**, the pre and postamplifier gains were left unchanged. In both figures, the pre- and postamplifier gains were the same. The vertical sensitivity of Channels 3 and A was also kept unchanged at 100 and 14.6 mV/div, respectively. A comparison of the amplitudes of the AE signals of Channel 3 shows that in both figures, when the abnormal pulse occurred, the AE signal amplitude was reduced considerably (**Figure 6.4B**) compared to the one obtained during normal operation (**Figure 6.3A**). The peak-to-peak amplitude during the abnormal operation was reduced to approximately 25% of the amplitude obtained in normal operation. The FFT transform in Channel A shows a similar reduction.

The data in **Figure 6.5A** and **B** were obtained using the R-50 AE transducer, coupled to the magnetron anode. The pre- and the postamplifier gains were held constant at 20 and 0 dB, respectively. The pulse rate was set to 1000 pps and the

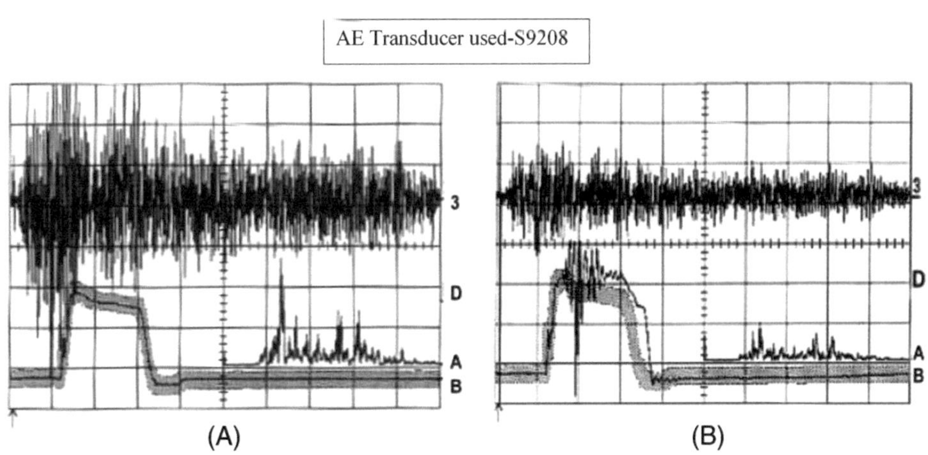

Figure 6.4.
(A) Normal operation. Pulse rate = 100 pps; normal pulse voltage = 15 kV; oscillator filament primary voltage = 115 V kept constant. (B) Abnormal operation. Pulse rate = 100 pps; pulse voltage increased to 19.5 kV. Ch 3: AE signal; Ch B: zoom of current pulse; Ch D: mask; Ch A: magnitude of FFT of Ch3. Preamplifier gain = 20 db and postamplifier gain = 0 db. Power output = 4.3 W. Ch 3: vertical sensitivity = 100 mV/Div and horizontal time scale = 0.1 ms/Div. Ch B: vertical sensitivity = 9 V/Div and horizontal time scale = 0.5 μs/Div. Observation: increased pulse voltage caused abnormal functioning leading to reduction in AE signal amplitude on Ch 3 in **Figure 6.4B** *(after Joshi et al. [2]).*

AE Transducer used-R50

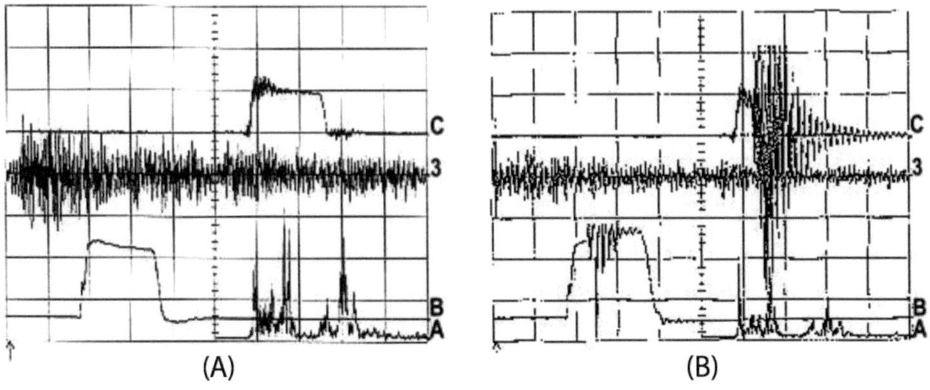

Figure 6.5.
(A) Normal operation. Pulse rate = 1000 pps; normal pulse voltage = 15 kV. (B) Abnormal operation. Pulse rate = 1000 pps; normal pulse voltage = 15 kV; oscillator filament primary voltage = 115 V but reduced to 95 V to stress tube. Ch 3: AE signal; Ch B: zoom of current pulse; Ch C: zoom of RF pulse; Ch A: magnitude of FFT of Ch 3. Preamplifier gain = 20 db and postamplifier gain = 0 db. Power output = 40 W. Ch 3: vertical sensitivity = 1 V/Div and horizontal time scale = 0.1 ms/Div. Ch B: vertical sensitivity = 13 V/Div and horizontal time scale = 0.5 µs/Div. Ch C: vertical sensitivity = 20 mV/Div and horizontal time scale = 0.5 µs/Div. Observation: reduced filament voltage caused abnormal functioning leading to reduction in AE signal amplitude on Ch 3 of **Figure 6.5B** *(after Joshi et al. [2]).*

pulse voltage to 15 kV. None of the microwave power of about 40 W flowing to the load at the end of the waveguide was reflected back.

In this test, all of these parameters were held constant at their normal values except for the oscillator primary filament voltage. **Figure 6.5A** shows the results when it was set at its normal value of 115 V, while **Figure 6.5B** shows the results when it was gradually decreased to a value of 95 V, when the abnormal current pulse occurred. As before, the display from the oscilloscope screen was automatically recorded on the disk.

In both **Figure 6.5A** and **B**, Channel 3 was connected to the output of the AE postamplifier. Channel B in **Figure 6.5A** shows the zoom display of the normal current pulse, while in **Figure 6.5B**, it shows the zoom of the abnormal current pulse. Channel C in **Figure 6.5A** shows the zoom display of the normal RF pulse, while **Figure 6.5B** shows the zoom of the abnormal RF signal. Although the stressing mode employed here was different from the previous one in **Figure 6.4B**, decreasing the filament voltage instead of increasing the pulse voltage, produced a similar decrease in the amplitude of the burst type of AE signal. Channel A shows the magnitude of the FFT of the AE signal.

The data in **Figure 6.6A** and **B** were obtained using the R-80 AE transducer. The pulse rate was 1000 pps and the pulse voltage was 15 kV. The oscillator primary filament voltage was set at the normal value of 115 V. In this case, a third parameter was used to induce abnormal functioning of the magnetron. First, data were obtained under normal parameter values when all the RF power was delivered to the matching load attached to the waveguide, as shown in **Figure 6.6A**. Abnormal functioning was then induced by introducing a load mismatch in the waveguide, between the magnetron and load, causing the power ratio of the reflected power to the incident power to increase, producing an abnormal current pulse. As before, the display on the oscilloscope screen was recorded on the oscilloscope's disk. As in previous tests, in both **Figure 6.6A** and **B**, Channel 3 was connected to the output of the AE postamplifier. Channel B in **Figure 6.6A** shows the zoom display of the normal

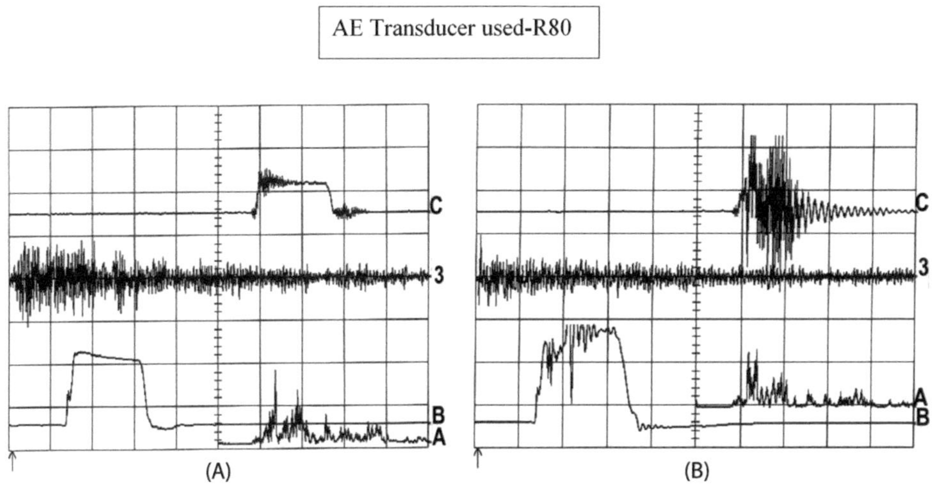

Figure 6.6.
(A) Normal operation. Normal pulse rate = 1000 pps; normal pulse voltage = 15 kV. (B) Abnormal operation normal pulse rate = 1000 pps; normal pulse voltage = 15 kV; oscillator filament primary voltage = 115 V kept constant. Ch 3: AE signal; Ch B: zoom of current pulse; Ch C: zoom of RF pulse; Ch A: magnitude of FFT of Ch 3. Preamplifier gain = 20 db and postamplifier gain = 0 db. Power output = 36 W forward but changed due to load mismatch in waveguide. Ch 3: vertical sensitivity = 1 V/Div and horizontal time scale = 0.1 ms/Div. Ch B: vertical sensitivity = 13 V/Div and horizontal time scale = 0.5 μs/Div. Ch C: vertical sensitivity = 25 mV/Div and horizontal time scale = 0.5 μs/Div. Observation: waveguide load mismatch reflected RF power back to tube causing abnormal functioning leading to reduction in AE signal amplitude in **Figure 6.6B** *(after Joshi et al. [2]).*

current pulse, while in **Figure 6.6B**, it shows the zoom of the abnormal current pulse. Channel C in **Figure 6.6A** shows the zoom display of the normal RF signal related to the RF power, while in **Figure 6.6B**, it shows the zoom of the abnormal RF signal. In **Figure 6.6B**, the amplitude of the AE signal showed a decrease due to the abnormal functioning of the magnetron, which was similar to that observed in the two earlier cases. Here again, irrespective of the cause of abnormal functioning, it produced a decrease in the amplitude of the AE signal. Channel A shows the magnitude of the FFT of the AE signal.

The experiment was repeated with the new S-9208A AE transducer. The suffix-A indicates that the transducer S-9208A carried a concave aluminum piece in front of the transducer faceplate to match with the convex surface of the magnetron cylindrical anode while the transducer S-9208 did not have the matching piece. The matching piece assured slightly better coupling. All parameters were initially set at normal values (filament primary at 115 V, the pulse voltage at 15 kV, pulse repetition rate at 1000 pps with all RF power delivered to the matched load), and the data for the normal functioning was obtained as shown in **Figure 6.7A**. Next the oscillator filament primary voltage was reduced gradually to 0 V from 115 V leading to occurrence of abnormal current pulse. The data of **Figure 6.7B** were then recorded immediately. All the channels show displays similar to the ones described above. In this case also, the amplitude of the AE signal on Channel 3 of **Figure 6.7B** decreased during the abnormal functioning of the magnetron tube relative to the AE signal of **Figure 6.7A** obtained with normal parameters.

The experiment was then repeated with the B1025 transducer. Here also, the abnormal functioning of the magnetron tube was induced by decreasing the oscillator filament primary voltage to 0 V and then increasing it back to 115 V. All other parameters were held at normal values (the pulse voltage at 15 kV, pulse repetition rate at 1000 pps with no RF power reflected back to the magnetron from the load). Once again, the amplitude of AE signal on Channel 3, shown in **Figure 6.8B**, obtained under abnormal functioning of the magnetron, showed a decrease in amplitude

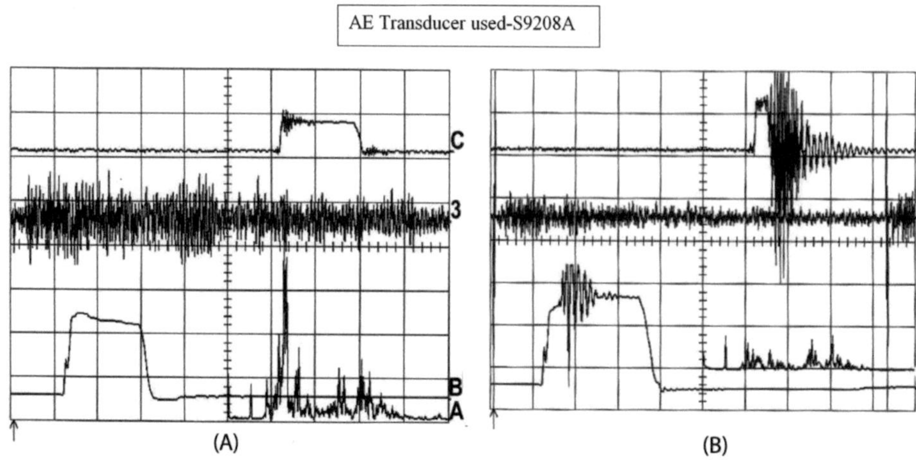

Figure 6.7.
(A) Normal operation. Normal pulse rate = 1000 pps; normal pulse voltage = 15 kV. (B) Abnormal operation. Normal pulse rate = 1000 pps; normal pulse voltage = 15 kV; oscillator filament primary voltage = 115 V but reduced to 0 V to stress tube. Ch 3: AE signal; Ch B: zoom of current pulse; Ch C: zoom of RF pulse; Ch A: magnitude of FFT of Ch 3. Preamplifier gain = 20 db and postamplifier gain = 0 db. Power output = 40 W. Ch 3: vertical sensitivity = 200 mV/Div and horizontal time scale = 0.1 ms/Div. Ch B: vertical sensitivity = 13 V/Div and horizontal time scale = 0.5 μs/Div. Ch C: vertical sensitivity = 25 mV/Div and horizontal time scale = 0.5 μs/Div. Observation: zero filament voltage caused abnormal functioning leading to reduction in AE signal amplitude on Ch 3 of **Figure 6.7B** *(after Joshi et al. [2]).*

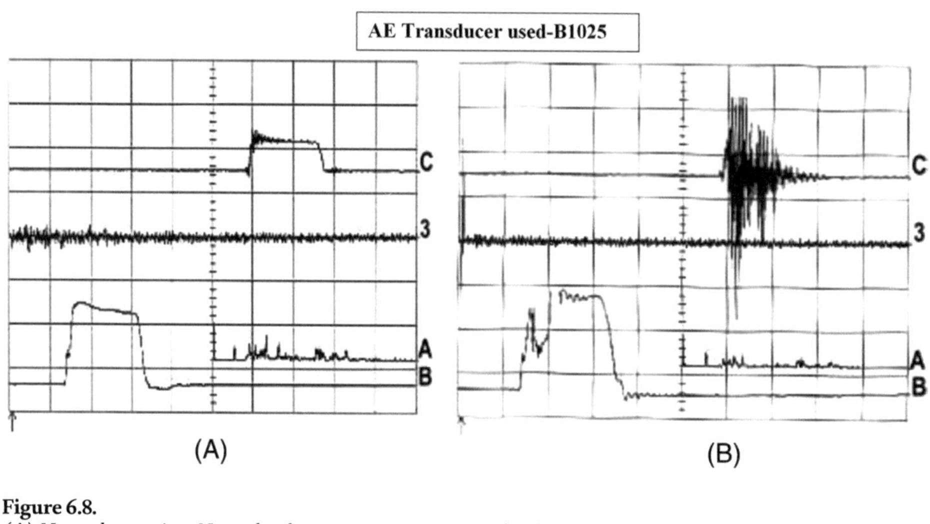

Figure 6.8.
(A) Normal operation. Normal pulse rate = 1000 pps; normal pulse voltage = 15 kV. (B) Abnormal operation. Normal pulse rate = 1000 pps; normal pulse voltage = 15 kV; filament voltage = 115 V but reduced to 0 V and backed up to 115 V to stress tube. Ch 3: AE signal; Ch B: zoom of current pulse; Ch C: zoom of RF pulse; Ch A: magnitude of FFT of Ch 3. Preamplifier gain = 20 db and postamplifier gain = 0 db. Power output = 40 W. Ch 3: vertical sensitivity = 500 mV/Div and horizontal time scale = 0.1 ms/Div. Ch B: vertical sensitivity = 13 V/Div and horizontal time scale = 0.5 μs/Div. Ch C: vertical sensitivity = 25 mV/Div and horizontal time scale = 0.5 μs/Div. Observation: cycling filament primary voltage caused abnormal functioning leading to reduction in AE signal amplitude on Ch 3 of **Figure 6.8B**.

relative to that of AE signal, shown in **Figure 6.8A** obtained under functioning of the tube.

During abnormal functioning of the high-power microwave tube magnetron caused by these three different stressing mechanisms in turn, the AE signals consistently showed a substantial decrease in amplitude relative to the AE signal amplitude recorded during normal functioning of the tube. The phenomenon observed was also independent of the AE sensors used to pick up the AE signals from the anode of the magnetron tube. The details of the mechanism of generating AE caused by the pulsing of the microwave frequency electromagnetic waves are under investigation. Currently, the pulsing electromagnetic energy generates thermal shock waves in the anode that in turn produce acoustic emission stress waves caused by the elastic properties of its material. The decrease in the pulsating electromagnetic energy in the abnormal functioning of the tube causes a corresponding decrease in the thermal shock wave energy and a subsequent decrease in the energy of the acoustic emission signal. Therefore, we can say that the current research work has proved experimentally the viability of the advanced nondestructive testing of acoustic emission as a noninvasive passive technique suitable for monitoring normal and abnormal functioning of high-power magnetrons. The AE technique does not have the limitations of the microprocessor-based technique discussed in the previous chapter, and the results are instantly available for any pulse rate ranging from 100 to 1000 pps. The next step is to extend this AE technique to the monitoring of other types of high-power microwave tubes such as klystrons and traveling wave tubes.

Acknowledgements

Figures 6.1 and 6.4–6.7 are reprinted with the permission of the American Society for Nondestructive Testing, Inc.

References

[1] Joshi NR, Brock DW, Russell SD, Kasa SD, Garcia GA. Acoustic emission technology for radar tubes. In: Technical Report 1810. San Diego, CA: SSC San Diego; 2000

[2] Joshi NR, Russell SD, Ramirez AD, Brock DW, Lasher ME. Acoustic emission technology for performance monitoring of high power microwave tubes. Materials Evaluation. 2007;**65**:411-416

Chapter 7

Acoustic Emissions from Traveling-Wave Tubes

7.1. Traveling-wave tube (TWT)

A TWT is an electron tube used for amplification at microwave frequencies. They typically operate over the frequency range above 500 MHz. The operation of a TWT depends on the interaction of a beam of electrons with an electromagnetic wave propagating on a slow-wave structure. The most common traveling-wave thermionic device is the linear-beam traveling-wave tube. Because of their very wide bandwidth, and high power gain, TWTs are extensively used in radar, space communication, and electronic countermeasure (ECM) systems.

A high-power TWT is a microwave amplifier characterized by high gain, wide bandwidth, relatively low efficiency, and high operating voltages. The electron gun assembly for a TWT consists of a cathode with a heating filament, a grid, an anode, and a collector. It is a form of tetrode tube. In addition, it has a slow-wave helix structure for amplification of radiofrequency (RF) signals. Electrons leaving the electron gun are accelerated through the helix structure to the collector by high voltages from the power supply feeding the electron gun of the TWT. **Figure 7.1** shows a schematic of a TWT with its power supplies [1]. A TWT uses a series of permanent magnets to provide a partial focus for the electron beam. The grid, anode, helix structure, and collector voltages "fine-tune" the beam focus so that a minimum of electrons strike the helix. Consequently, very few electrons strike the helix structure. An RF signal passes into the interior of the tube through a ceramic window or coaxial connector at the RF input and then propagates along the slow-wave circuit (the helix) with a phase velocity approximately equal to the electron beam velocity. The beam and the propagating RF signal interact such that energy is transferred to the RF signal on the helix. An RF signal propagating on the helix structure causes alternately accelerating and decelerating electric fields along the helix structure. Electrons passing through the helix will either speed up or slow down, depending on where they are in the field. Electrons that are accelerated meet with electrons that are slowed down by the preceding field. Thus, the input RF signal interacts with the electron beam causing it to bunch up. The bunched electron beam then generates its own alternating electric fields that interact with the electrons in the helix, increasing the amplitude of the RF signal on the helix. This process continues as the beam travels through the helix and the amplitude of the RF wave associated with the bunched electron increases. The amplified RF signal then passes out of the interior of the TWT at the RF output ceramic window or coaxial connector, and the collector then absorbs the spent electron beam. The resulting heat energy is dissipated in the collector structure and TWT heat sink, which are cooled by a high-efficiency air blower. The helix has a virtually constant phase velocity over a wide range of frequencies and has by far the largest bandwidth of any slow-wave structure. This helps to give TWTs their wide operating bandwidths. It possesses a high interaction factor between the helix and electron beam and is thus able to produce a high gain in a short length [2].

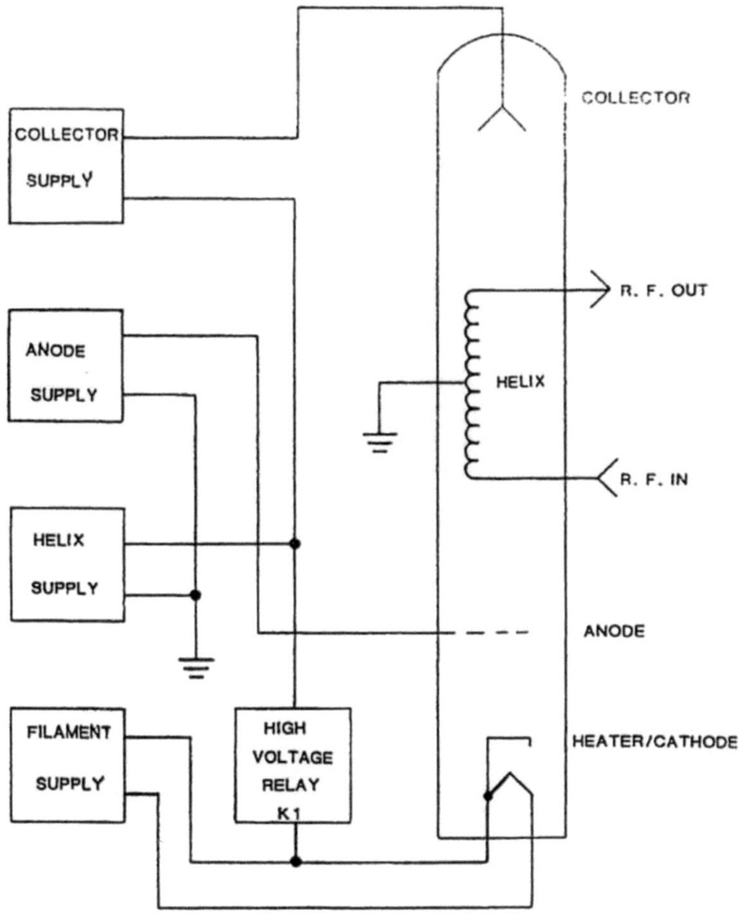

Figure 7.1.
Schematic of traveling-wave tube (tetrode) (after Logimetrics [1]).

7.2. Traveling-wave tube amplifier (TWTA)

The TWTA (Model A600/L-878A) used in this research work was designed and manufactured by Logimetrics, Inc. [1]. It was designed to linearly amplify microwave signals over an operating bandwidth of 1 GHz (1–2 GHz). The TWTA unit was self-contained. It obtained its power from a three-prong wall socket through a cable rated at 30 A/115 V. The intake air passed through a filter on the front panel. An exhaust fan was located at the rear of the unit. The front panel had three meters: the collector current (range: 0–1 DC A), the helix current (range: 0–100 DC mA), and the forward and reverse RF power output (range: 0–500 W forward and 0–100 W reverse). The front panel also included an amplifier RF gain control, which was a 24.5 turn potentiometer that gave 0–20 dB of attenuation. The unit was protected by five overload shutoff mechanisms. The front panel also had a set of five fault indicator lamps, indicating TWT thermal overload, power supply thermal overload, cooling air supply fault, VSWR (voltage standing wave ratio) cutoff, and helix current fault. The unit had an AC power main breaker, power-on switch, and a two-position standby/RF-on switch. A timer inside the unit provided a 3-min wait, after the main circuit breaker and "power on" switch were turned on. When the standby indicator lamp turned off, the switch could be changed to its "RF-on" position.

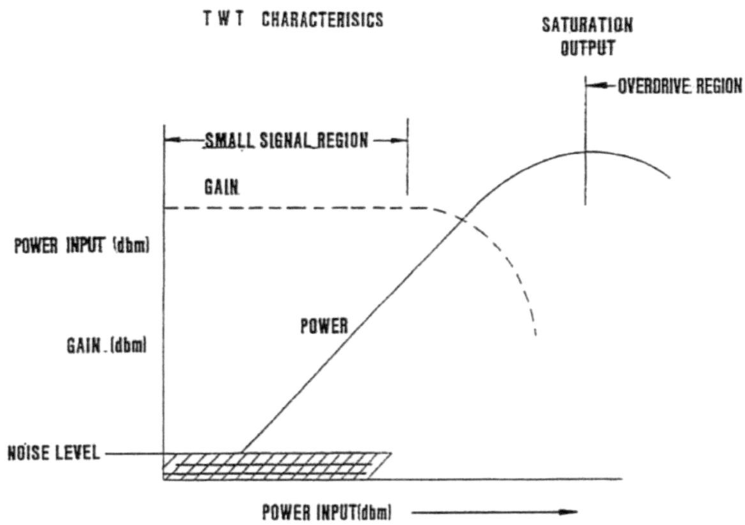

Figure 7.2.
Input-output characteristics of TWTA (after Logimetrics [1]).

The TWTA used in this work provided more than 200 W of CW (continuous-wave) output power over its operating range of 1–2 GHz. The unit was protected by helix current- and voltage-sensing circuitry. An internal RF sensor system was designed to detect high VSWR and automatically turn the amplifier off while simultaneously lighting up a VSWR fault lamp on the front panel. The input and output impedances of the TWTA were 50 ohms, but it could operate safely into an impedance mismatch that produced a VSWR of 2.5:1. The input and output connectors were "N" type.

The input/output characteristics of a typical TWTA are shown in **Figure 7.2**. The small signal region is characterized by linear operation, low intermodulation products, and low harmonic generation. The gain of a TWTA is usually 6–8 dB lower at saturation than the gain available for small signals. The gain decreases further in the overdrive region [1].

7.3. Experimental setup

Figure 7.3 shows a schematic drawing of the experimental equipment and connections. In normal operation of the TWTA, the output is connected to a waveguide or coaxial cable, leading to an antenna. In the laboratory, it was connected through a high-power directional coupler to a dummy load rated at 50 ohms and 225 W. To remove heat, the dummy load was equipped with a fan that could be run directly from the regular power supply. The high-power directional coupler was rated for 200 W over a frequency range of 960–1215 MHz, with a coupling of 30 dB. It had four "N"-type connectors J1, J2, J3, and J4. The input connector J1 was connected through an "N"-type coaxial cable (Alpha Wire 92194-RG 214/U) to the output terminal on the rear panel of the TWTA. The output connector J2 was connected through an "N"-type coaxial cable to the dummy load. The forward connector J3 was connected through an "N"-type coaxial cable and then through 20- and 10-dB attenuators to the input terminal (50 ohm, 9 MHz–22 GHz) of an HP 8562A Spectrum Analyzer. The two attenuators had medium power ratings of 20 W and frequency ranges of DC-11 GHz. They had metal fins to dissipate heat. These two attenuators

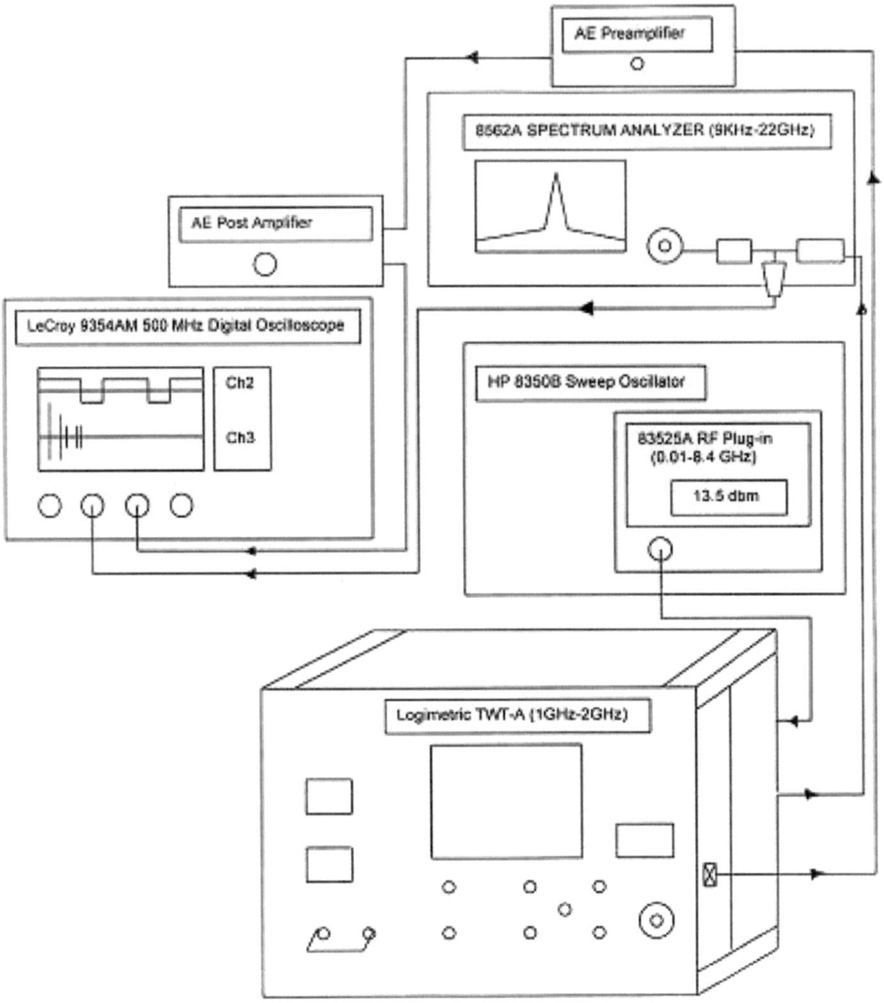

Figure 7.3.
Schematic of AE experimental equipment with TWTA (after Joshi et al. [3]).

are visible in **Figure 7.3**, in front of the image of the 8562A Spectrum Analyzer, with the 10-dB attenuator on the left and the 20-dB attenuator on the right.

Figure 7.3 also shows a "T" connector between the two attenuators, which is connected to an inline crystal detector with an "N"-type connector on one end and a BNC connector on the other end. It was a silicon microwave diode square law crystal detector. Its output was connected through a regular BNC cable (Pomona Electronics 2249-C-48) to Channel 2 of a LeCroy 9354AM 500 MHz Digital Oscilloscope. The input signal to the TWTA was generated by an HP 8352A RF plug-in module installed in an HP 8350B Sweep Oscillator. The "N"-type output connector from the oscillator module was connected through an "N"-type coaxial cable to the "N"-type input connector on the rear panel of the TWTA [1].

The right cover of the rectangular housing of the TWTA unit was then opened. The TWT tube enclosed in the metal box was located under the air circulation duct. An acoustic emission (AE) transducer (R50-Physical Acoustics) was then mounted on one side of the TWT metal casing using a silicone compound (Mil-S-8660B, General Electric Company) as a couplant. It was held in position by a padding of foam inserted between the back of the transducer and the frame of the TWTA unit. The transducer was then connected to the preamplifier, with the preamplifier output connected, through the AE1A postamplifier, to Channel 3 of the LeCroy Digital Oscilloscope.

The sampling rate of the oscilloscope for a single channel was 2 GS/s, with a data storage capability of 1 million points. If all four channels were used at the same time, the sampling rate was 500 Ms/s with 250 kilopoints per channel.

The R50 transducer was a general-purpose transducer useful for monitoring AE signals over the range of 100–700 kHz. The 2/4/6 preamplifier had three gain settings of 20, 40, and 60 dB, respectively. It had a high input impedance. It could produce large output signals up to 20 V peak-to-peak into 50 ohms. The useful frequency range of the preamplifier was 100 kHz to 1 MHz. The preamplifier was enclosed in a metalized bag to prevent it from picking up extraneous signals. The AE postamplifier had a total gain of 41 dB, achievable in 3-dB steps. The input and output impedances at the terminals "AE INPUT" and "AE OUTPUT" were 50 ohms. The useful frequency range of the postamplifier was 100 kHz–1 MHz. Other details about the experimental setup can be obtained from Joshi et al. [3].

7.4. Experimental procedure

The maximum input power recommended for the spectrum analyzer is 0 dBm or 1 mW. The input power from the 83525A RF plug-in could be changed in steps. The frequency of the CW input signal to the TWTA was initially chosen as 1.2 GHz. The output of the TWTA unit was connected to the 50-ohm dummy load through the high-power directional coupler. The RF gain control knob (0- to 20-dB attenuation) on the front panel was set all the way to the clockwise position, indicating that the full input power was applied to the TWTA for amplification. The front panel power indicator had two scales, for the forward and reverse power levels, corresponding to the up and down positions of the selector switch. The output power was attenuated by 60 dB (30-dB directional coupler +20-dB attenuator +10-dB attenuator) before it was fed to the input terminal of the spectrum analyzer.

7.5. Experiments and results

7.5.1 Part I (CW mode with dummy load of 50 ohms)

In this initial test, the aim was to observe acoustic emission activity under normal operation of the TWTA in the CW mode when its output was connected to the 50-ohm dummy load through the 30-dB directional coupler. The AE transducer was mounted on the TWTA near the output terminal of the helix. The normal noise band on Channel 3 was superimposed onto a mask (shown in **Figure 7.4**, Zoom Trace D) with sufficient width to avoid triggering on the noise signal. The trigger level was chosen to be slightly below the upper boundary of the mask. The trigger position was set to two divisions to the right of zero on the horizontal scale. If all the points of the signal remained within the mask, there was no triggering of the oscilloscope to dump its screen display onto the disc. This arrangement ensured the recording of AE burst signals only and avoided noise signals. To avoid any damage to the oscilloscope screen caused by long display times, the mask on the Zoom Trace D was set and its display was then turned off during prolonged operation of the equipment. Nine AE events were recorded on the disc in 3 hours. It was expected that changes would be seen in the detected RF level on Channel 2 whenever AE signals were produced, because of the irregular functioning of the TWTA. **Figure 7.4** shows a typical AE signal on Channel 3, with the voltage from the RF level detector

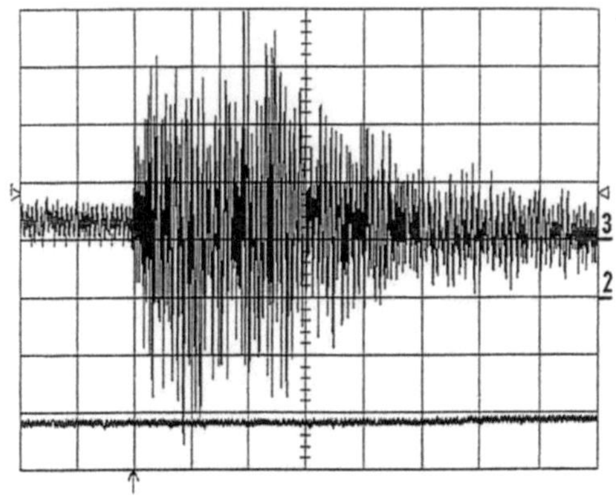

Figure 7.4.
AE signal from CW TWTA. Ch 3: Sporadic AE signal. Ch 2: Detected RF signal at −22 mV. Waveguide terminated with 50 ohms. Chs 2 and 3: Horizontal scale = 0.1 ms/div. Ch 2: Vertical sensitivity = 10 mV/div. Ch 3: Vertical sensitivity = 50 mV/div (after Joshi et al. [3]).

into Channel 2 at −22 mV. Instead, no appreciable changes were detected in the AE signal or in the RF level during the 3 hours.

7.5.2 Part II (CW mode with no load-all power reflected back)

7.5.2.1 Effects of closing the front air inlet

The objective of this second test was to stress the TWTA's operation by forcing it to operate at higher temperatures, which in turn might produce irregular RF-amplified pulses and corresponding changes in the AE signals. The front panel air inlet was covered tightly with aluminum foil sealed around its edges. The dummy load at the RF output was disconnected and replaced with a Teflon (insulator) plug between the central pin and the surrounding metal cylinder of the "N"-type connector. This was done to reflect the RF power back into the TWTA to stress the tube. The TWTA was then put back into operation, and AE signals were received at intervals. After 8 hours, the TWTA shut itself off due to thermal overload. Compared to the first test, there was no change in the detected RF level on Channel 2. Also, no changes in AE signals on Channel 3 were detected during the operation of the tube at higher temperatures.

7.5.2.2 Effects of closing the front air inlet and the rear air exhaust

The next step in trying to induce irregular functioning of the TWTA was to close both the front panel air inlet and rear panel air outlet with aluminum foil sealed at the edges. Additionally, all power was reflected back, with a Teflon plug replacing the dummy load. As before, the input RF power was 13 dBm. The forward RF output power was 120 W, and the reverse power was greater than 100 W. The unit shut off in 10 min due to the thermal overload. The copper plate, mounted onto the TWT, and acting as its heat sink, became very hot. It had a heat sensor that then signaled the unit to turn off. Temperature measurements were taken by a thermocouple (Omega Engineering Corporation) pressed against the copper plate heat sink. The temperature rose to 83°C in 9 min. At this stage, it was decided to free the air outlet on the rear panel of the unit. In the repeated experiment with both air passages blocked, the temperature rose to 104°C in 12 min. However, all these tests did not produce any irregular RF output pulses and corresponding signature AE signals. At this stage, the left side panel and the top side panel were opened. The five sensors protecting the unit were identified. They were (1) thermal sensor, (2) air flow sensor, (3) power supply sensor, (4) VSWR overload sensor, and (5) helix current sensor.

7.5.2.3 Effects of different locations of the AE sensor

The next step was to find out whether the position of the AE transducer could make any difference in receiving the AE signals. The transducer was mounted on the metal coaxial cable that ran from the helix output terminal to the output terminal on the rear panel. The new position of the transducer made no difference on the recorded AE activity. The AE activity was sporadic, on-and-off, and only appeared intermittently. A more serious way to stress the unit was to disconnect the thermal sensor and close both the air inlet and air exhaust outlet. However, it was decided not to follow that course, but instead to feed a pulsed RF input signal to the TWTA in place of the CW signal and observe the effects.

7.5.3 Part III (pulse mode with no load: all power reflected back)

In this portion of the experiment, the RF input frequency was initially chosen as 1.4770 GHz. The frequency span (Δf) was set to zero. The sweep was triggered from the 60-Hz power line. The "RF on" time was set for 2 s. The "RF off"' time was fixed internally at about 1 s. On the spectrum analyzer display screen, the start and stop frequencies were set to 1 and 2 GHz. The resolution and video bandwidths were both set to 300 kHz. The sweep time was set to 50 ms. The RF plug-in module, generating the input signal, was set for a power level of 13.5 dBm. The ALC (automatic level control) mode was set to the internal sensor. The TWTA was then turned on. After a few minutes, the power meter on the TWTA front panel was showing rises and falls in the output power in sync with the rises and falls of the RF signal on the spectrum analyzer screen.

The input frequency was then varied, and the maximum power output changed with changes in the input frequency. By gradually changing the input frequency, the output power peaked at around 1.5040 GHz. The output power at this frequency was 140 W. As mentioned earlier, the AE transducer was mounted on the TWT tube near the helix output terminal. The AE transducer preamplifier and postamplifier gains were 40 and 21 dB, respectively. With this setup, AE signals were generated in step with the oscillating output power of the TWTA. They are not sporadic as before.

Note: In **Figure 7.6**, the AE signal appears compressed on the sixth vertical division between the left lower ON (2 s) segment and right upper OFF (0.7 s) segment of RF.

This result may be due to the pulsed RF power producing intermittent heating. The subsequent contraction during the interpulse cooling period could be the cause of the observed AE energy. **Figure 7.5** shows the AE signal on Channel 3 generated at the transition from the "RF on" state to the "RF off" state as displayed on Channel 2. The AE signal appears as a very narrow vertical line, close to the sixth vertical division, due to the choice of a horizontal time scale suited to show "RF on" and "RF off" segments, which are on a longer time scale. The "RF on" and "RF off" segments are inverted in Channel 2 of the figure, due to the negative voltage output from the pulses.

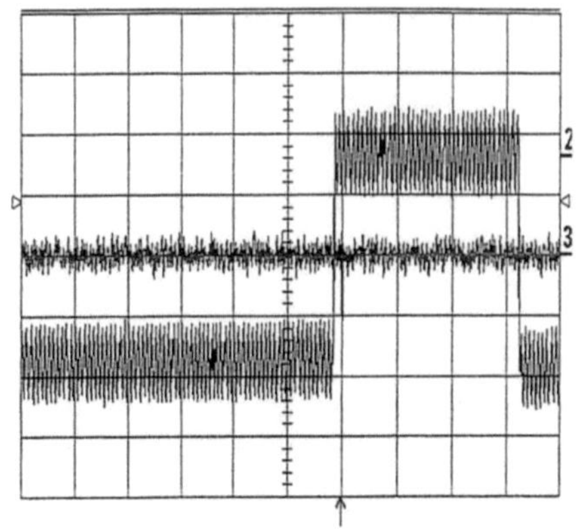

Figure 7.5.
AE signal from pulsed TWTA. Ch 3: Periodic AE signal in sync with pulsing TWTA on the sixth vertical division. Ch 2: Detected pulsed RF signal. Chs 2 and 3: Horizontal scale = 0.2 s/div. Ch 2: Vertical sensitivity = 10 mV/div. Ch 3: Vertical sensitivity = 50 mV/div (after Joshi et al. [3]).

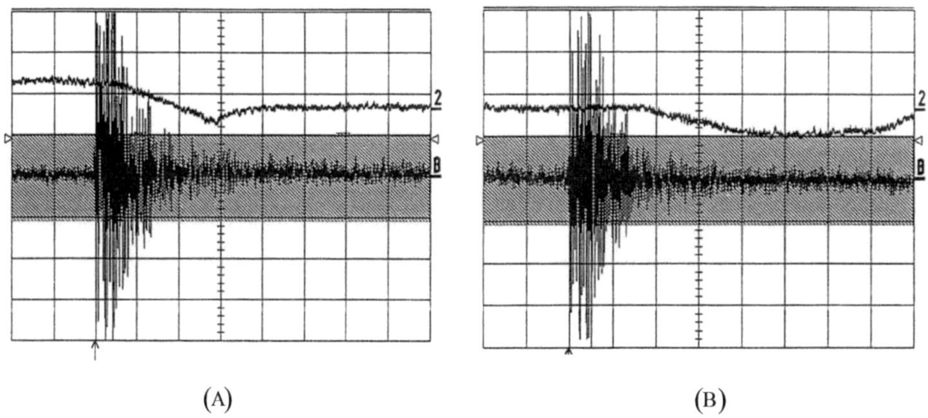

Figure 7.6.
(A) AE signal from pulsating RF input to TWT-A. Ch2: Pulsating RF voltage at −22 mV. level. Zoom trace D: Mask in action. Ch 3: AE signal from pulsating input. Zoom trace B: Zoom of Ch 2. Ch 3: Time scale = 0.5 ms/div and vertical sensitivity = 50 mV/div. Ch 2: Time scale = 0.5 ms/div and vertical sensitivity = 10 mV/div. RF input power 13.5 dBm at frequency = 1.5041 GHz and RF output power (forward) = 140 W and RF output (reversed) = 100 W. No matching load-all RF power was reflected back. (B) AE signal from pulsating RF input to TWT-A. Ch2: RF voltage is changing its value from that of **Figure 7.6A**. *Zoom trace D: Mask in action. Ch 3: AE signal from pulsating input. Zoom trace B: Zoom of Ch 2. Ch 3: Time scale = 0.5 ms/div and vertical sensitivity = 50 mV/div. Ch 2: Time scale = 0.5 ms/div and vertical sensitivity = 10 mV/div. RF input power 13.5 dBm at frequency = 1.5041 GHz and RF output power (forward) = 140 W and RF output (reversed) = 100 W. No matching load-all RF power was reflected back (after Joshi et al. [3]).*

7.5.3.1 Effects of RF input duty cycle on AE activity

The next part of the test series was to examine the effect of duty cycle of the RF pulses on the generation of AE signals. The "RF on" interval was increased from 2 to 20 s in 1-s increments. The "RF off" interval was fixed internally at about 1 s. In every instance, AE activity was observed in step with the changing RF output. The "RF on" interval was then reduced to 1 s, almost equal to the "RF off" interval. The AE signals were still generated in step with the changing RF output. The "RF on" interval was then further reduced to less than 1 s, and the AE activity diminished rapidly with further decreases in the "RF on" time. It appeared that the "RF on" interval had to be greater than the "RF off" interval to generate AE activity. It was decided, therefore, to

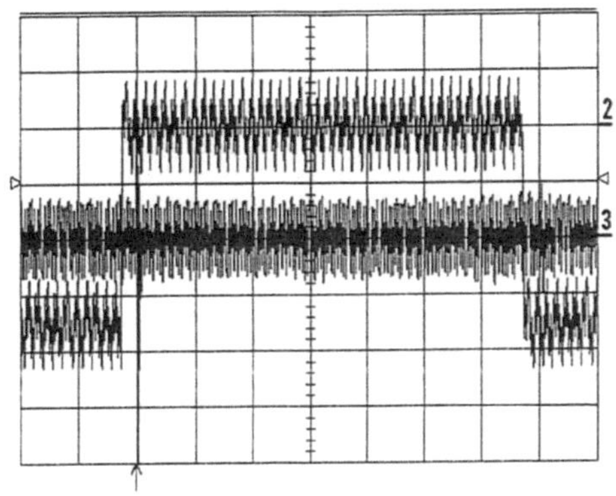

Figure 7.7.
TWT-A in pulsating mode as opposed to CW mode. Ch 3: AE signal on the second vertical division (compressed) in sync with pulsating TWT-A. Ch 2: Detected RF (inverted) at 35 mV. level. The top segment is RF OFF (0.7 s) and the lower half is the RF ON segment. Noise band on the central line is 15 mV. Ch 3: Time scale = 0.1 s/div and vertical sensitivity = 50 mV/div. Ch 2: Time scale =0.1 s/div and vertical sensitivity = 10 mV/div (after Joshi et al. [3]).

set the "RF on" interval to 2 s for further testing. **Figure 7.6A** and **B** shows AE signals generated with a pulse rate of 20 ppm. In **Figure 7.6B**, the detected RF level is at the end of a pulse and is in the process of decreasing its value from that in **Figure 7.6A**. For each pulse, the "RF-on" interval was 2 s and the "RF off" interval was 1 s. In this test, all RF power was reflected back since the dummy load was replaced with the Teflon plug.

Next, to determine the "RF off" interval precisely, the horizontal time scale was changed and the screen displays were saved on the disc. In **Figure 7.7**, a horizontal time scale of 0.1 s/div was used. It was then observed that the "RF off" interval was 0.7 s (Channel 2), slightly less than the assumed value of 1 s. The "RF off" interval was internally fixed and, hence, could not be changed. The AE signal from Channel 3 appears on the second vertical line.

7.5.3.2 Effects of sudden "cutoff" of the filament power on the performance of the TWTA

As mentioned before, the purpose of this research project was to examine nondestructively the irregular performance of the TWTA. As described earlier, the TWTA unit used in the project was protected by five overload detection sensors. Consequently, one has to use novel methods to induce irregular performance of the unit. We decided to suddenly cut off the power to the cathode heater filament and observe its effect on the performance of the tube. Since the cathode heater filament was at a high negative voltage of about −2000 V with respect to the collector, high-voltage protection equipment was needed during the sudden disconnection of the power supply to the filament. As the tube rapidly cooled off, many AE signals were generated in succession over a time interval of about 1 min before all the acoustic emissions stopped as the forward power output dropped rapidly from 140 to 0 W. Since it took 15 s for the oscilloscope to store the screen display onto its disc, it was impossible to save all of the AE signals on the floppy disc, as they were generated during the rapid cool-off period. However, a couple of screen displays were stored automatically during this interval. The AE signals obtained during this cathode cooling experiment had relatively small amplitudes and longer ringing time.

7.5.3.3 Effect of cold start

There was a standby time of 3 min between setting the switch to the "power on" position and being able to engage it to the "RF on" position. After closing both of the inlet and outlet windows for the air, the standby time was measured. It turned out to be the same as before at about 3 min. Again, in an attempt to produce irregularities in the functioning of the unit, the timing circuit governing the standby interval was bypassed by changing the circuitry. The input RF power was set at 13.5 dBm, as before. The RF input power was pulsed with an "RF on" interval of 2 s. No load was connected so that all of the RF power was reflected back. The main circuit breaker switch, the "power-on" switch, and the "RF on" switch were turned on in succession to produce a cold-start condition. No damage was done to the tube. The only observable effect of the cold start was a short delay in receiving AE signals, which were again in step with the pulsed RF power. The AE signals generated initially were not strong. After about 10 min, strong AE signals started appearing, which were again in step with the pulsed RF input and output.

7.5.4 Part IV (pulsed mode with 50-ohm dummy load)

In this segment of the project, a dummy load of 50 ohms was connected to the output cable of the TWTA, and the RF input to TWTA was set to the CW mode as in Part I of the experiment. Again, the RF input power was 13.5 dBm. The gain setting was set to the fully clockwise position, giving the maximum gain. There was no AE activity observed, even after half an hour. Then, the RF input signal was changed from the CW mode to the pulsed mode. The dummy load remained connected. Within a few minutes, AE signals were observed (**Figure 7.7**). They were in step with the pulsed RF output.

7.6. TWT amplifier operation conclusions

Operation in the CW mode did not produce AE activity in the two cases where the output cable of the TWTA unit was connected to the dummy load of 50 ohms (matched load) and then to the Teflon plug (mismatched load), which caused the RF power to be reflected back.

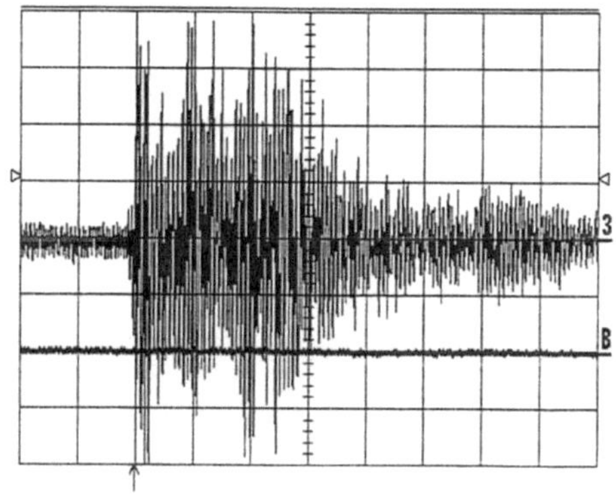

Figure 7.8.
AE signal from pulsating TWT-A input. Ch 3: Expanded AE signal. Zoom trace B of Ch 2: Detected RF at level of about 40 mV. Ch 3: Time scale = 0.1 ms/div and vertical sensitivity = 50 mV/div. Zoom trace B of Ch 2: Time scale = 0.1 ms/div and vertical sensitivity = 20 mV/div (after Joshi et al. [3]).

The pulsed RF input power generated AE activity, irrespective of the termination (matched or mismatched load) to the output cable.

This AE activity was generated as long as the "RF on" interval in the pulsed RF input power was longer than the "RF off" interval. AE signals were generated at the transition from the "RF on" segment to the "RF off" segment of the output RF pulse. The cold-start experiment showed that the strength of AE signals was low at the beginning but then picked up in intensity after the unit had been in operation with pulsed input power for about 10 min.

The simulated cold cathode experiment showed that rapid changes in the filament power supply voltage caused successive outputs of AE activity that dropped off to zero with the diminishing RF power output of the TWTA.

Overheating of the unit did not produce any new effects in AE activity because the thermal overload sensor protected the unit.

The maximum available leveled output from the 83525A RF plug-in, mounted in the 8350B sweep oscillator, was 17.5 dBm. However, the TWTA was shut off by its VSWR overload sensor when the input power exceeded 14 dBm.

The TWTA used in this project was designed for CW mode operation. By operating it in the pulsed mode, AE activity was produced in step with the changing RF input power. During operation in the CW mode whenever there was a change (increase or decrease) in the RF input power, AE activity was generated. In short, changes in the RF output power due either to irregular changes or to regular changes in the RF input power caused AE signals to appear. These experimental observations definitely indicate that AE activity will be produced whenever there is a change in the RF output power of the unit, irrespective of the cause of the change, either external to the unit (RF input power) or internal to the unit (degradation of the tube performance). It can safely be said that acoustic emission activity can be used as a nondestructive testing tool for in situ performance monitoring of normal and abnormal functioning of high-power microwave TWT amplifiers.

Acknowledgements

Figures 7.3–7.8 are reprinted with the permission of the American Society for Nondestructive Testing, Inc.

References

[1] Logimetrics. Traveling Wave Tube Amplifier (TWTA). In: Technical Manual, ModelA600//L-878A. Bohemia, NY, U.S.A.: Logimetrics, Inc; March 1999

[2] Hughes Aircraft Company. TWT and TWTA Handbook. Torrance, CA: Hughes Aircraft Company, Electron Dynamics Division; 1979

[3] Joshi NR, Brock DW, Russell SD, Lasher ME, Kasa SD. Built-in test for high-power microwave tubes using acoustic emission technology. In: Technical Note 1816. SSC San Diego; 2001

Chapter 8

Acoustic Emissions from Klystrons

8.1. What is a klystron?

The usefulness of the advanced nondestructive emissions (NDEs) technique of acoustic emission (AE) was proven in the performance monitoring of magnetrons and traveling wave tubes (TWTs). The next area of interest was the application of AE technology to the operation of klystrons. High-power microwave tubes may be divided into two general categories: crossed-field devices, characterized by orthogonal electric and magnetic fields, and linear-beam devices, having a continuous electron beam traversing an interaction region. Crossed-field devices are typified by magnetrons and crossed-field amplifiers, while linear devices include cavity klystrons, TWTs, twystrons, and extended interaction oscillators. Klystrons are available as both oscillators and amplifiers. They are inherently narrowband devices, although some have a degree of tunability. They have high gain and are capable of either pulsed or continuous wave (CW) operation at high peak and average power levels [1].

Klystrons have three sections: the electron gun and beam system, RF circuit, and beam collector system. Klystron radio-frequency (RF) circuits are fairly simple, at least when compared with traveling wave tube circuits. The RF circuit comprises a number of resonant cavities, typically between three and six. The electron beam passes axially through the centers of the cavities. A high voltage is applied between the cathode and collector. Electrons emitted from the cathode are accelerated through the long tube. At the tube's mouth, and throughout the tube's length, the electron beam is focused by an external magnetic field provided by sets of flat coils concentric with the axis of the tube. Each of the resonant cavities can be tuned by symmetrically altering its physical dimensions mechanically. The first cavity is excited by an externally generated low-level RF signal introduced by a coupling loop in the cavity [2].

The basis of klystron operation is the very common phenomenon that reduces the efficiency of conventional electronic tubes, which is the transit time. The field created across the beam by the cavity's oscillating electric field causes the electrons to either accelerate or slow down, depending upon the time when they pass through the cavity. The beam energy is therefore velocity modulated in phase with the oscillations of the first cavity. This very-high-frequency bunching (density modulation) is accentuated by the subsequent cavities that are tuned to the input frequency. The cavities are caused to resonate by the electric field of the bunched electron beam passing through their interaction space. As the beam progresses through the drift tube, along the axis of the cavities, through these successive interaction spaces, the amplitude of the beam bunching continues to grow and becomes much greater in magnitude than the amplitude caused by the original modulating input. At the final cavity, whose excitation is a few orders of magnitude greater than that in the first cavity, the RF output power is extracted, again by a coupling loop. The RF energy then leaves the klystron, through a coaxial cable or waveguide. Power gains of up to 40 dB are possible at typical radar frequencies. But the klystron's chief value as a radar transmitter is its ability to sustain high mean power levels in its amplifier role [3].

Some detuning of the intermediate cavities could increase the power output. The bandwidth of a klystron can be increased by loading its intermediate cavities, but the overall gain is reduced in the process. As in the case of a vacuum tube amplifier with a single tuned circuit in its output, the gain-bandwidth product tends to remain constant. Stagger tuning the intermediate cavities (tuning them slightly above and slightly below the input cavity frequency), along with loading the cavities to broaden their response, will give the klystron amplifier a broader bandwidth but with reduced power gain. To achieve a very wide bandwidth, a slow wave structure must be used. This would require a TWT design. The TWT has an inherently large bandwidth, but it cannot produce the high peak power outputs that multicavity klystrons can achieve. When pulsed, it is possible to obtain peak powers of between 0.5 and 20 MW at efficiencies of over 40% at frequencies near 3 GHz [4]. Klystrons are commonly used at ultra-high frequencies in CW mode for television transmission at power outputs of up to 60 kW. Continuous wave use implies that the mean and peak outputs are nearly the same. In short, the high-power pulsed klystron is an amplifier that is characterized by high gain, high peak power, and good efficiency. It has relatively narrow bandwidths and requires relatively high voltages. Its large physical size and the associated modulators and x-ray shielding have limited its applications [5].

8.2. Experiments on the first klystron unit

The unit used in this effort was a five-cavity klystron amplifier. The pulse repetition frequency (PRF) was set at 213 pps. Pulses were generated in pairs. The pulse width of the first pulse was 2 μs and that of the second pulse was 125 μs. The separation between two pulses was 125 μs. The RF frequency of the first pulse was 894.33 MHz. The second pulse was a chirp signal with a center frequency of 894.33 MHz and a variation of ±1/2 MHz, giving a chirp range of 1 MHz. The klystron had 48 channels between 851 and 942 MHz. The low band consisted of channels 1–16, the middle band consisted of channels 17–32, and the high band consisted of channels 33–48. The klystron was connected to the dummy load through a waveguide. The dummy load was cooled by circulating water. In normal operation, the cathode was biased at −42 kV for the grounded collector.

8.3. Experimental setup

The same AE analysis technology used on CFAs in Chapter 6 and on TWTs in Chapter 7 was applied to a klystron tube. This AE system included the general purpose R50 transducer, pre-amplifier, and post-amplifier, all described earlier. **Figure 8.1** shows the different locations on the klystron unit, where the AE transducer was attached to pick up emissions generated during klystron operation. The pre-amplifier was wrapped in a bag of aluminum foil to prevent it from picking up extraneous signals. Appropriate coaxial cables (RG 223 U) were used for all connections. The gains of the pre-amplifier and post-amplifier were set to 40 and 21 dB, respectively. As before, a LeCroy 9354 AM 500-MHz digital oscilloscope was used to observe the AE signals generated and to record them on a disc, if necessary. **Figure 8.2** shows schematic setup for the klystron. Channel 1 of the oscilloscope was connected to the input (RF drive) of the klystron amplifier. Channel 2 of the oscilloscope had a bayonet neill-concelman (BNE)-connector with one end connected to the beam current sensor connector (5 A/V) of the klystron unit and the other end to the channel 4 through a combination of a high-pass filter (cutoff, 1.67 MHz), detector, low-pass filter (cutoff, 15 MHz), and threshold amplifier. The threshold amplifier has its own amplifier.

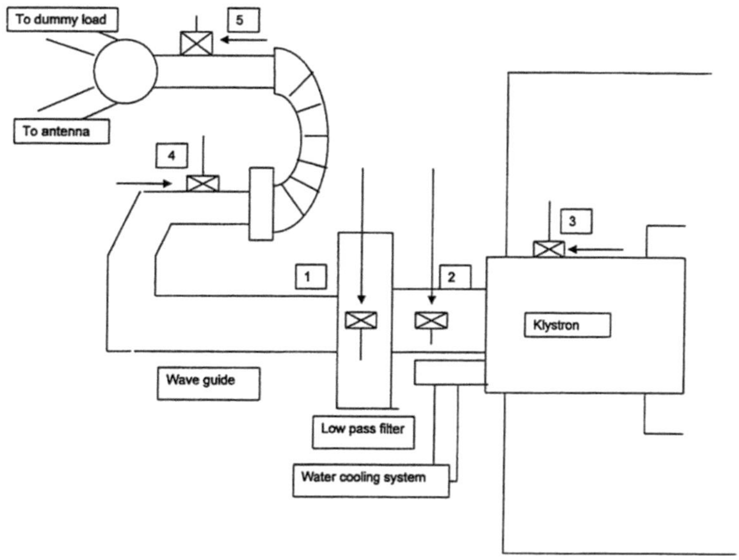

Figure 8.1.
Schematic diagram of AE transducer locations on the klystron radar unit (after Joshi et al. [7]).

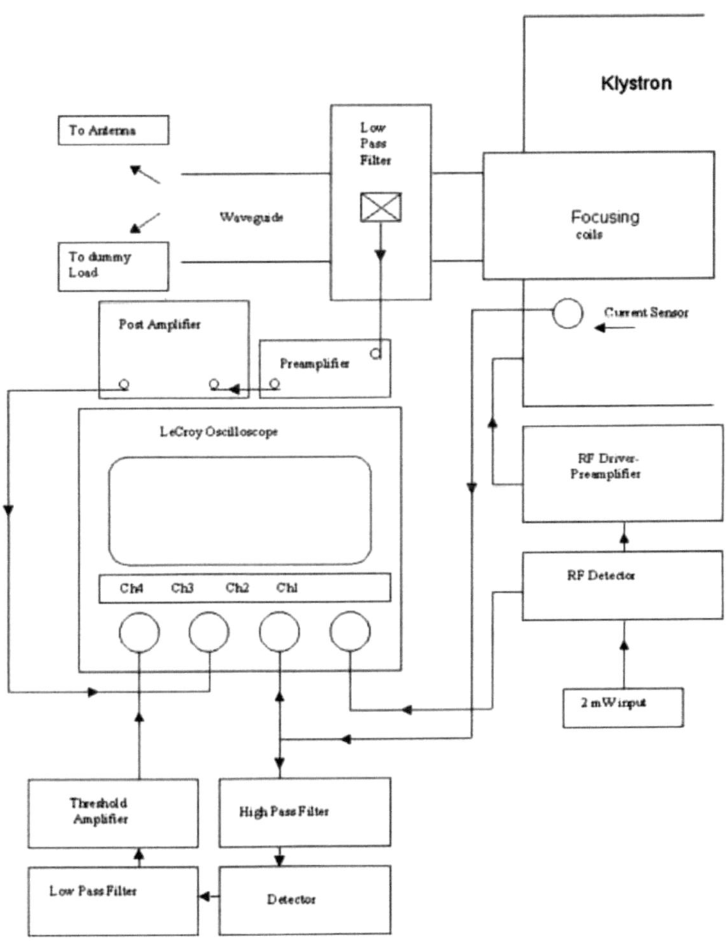

Figure 8.2.
Schematic diagram of experimental setup for AE testing of klystron (after Joshi et al. [7]).

8.4. Experimental results from the first klystron

8.4.1 Effect of different locations of AE sensor

When all the parameters, such as the gains of the two AE amplifiers and the transducer couplings, were held constant, the strength of the AE signal differed from location to location during normal operating conditions of the klystron unit. The AE signals received from Location 5 on the waveguide (**Figure 8.1**) and from Location 3 on the focusing coils inside the cabinet of the klystron unit were stronger than those received from Location 1 on the low-pass filter.

8.4.2 Effect of increased peak power output

The klystron unit was well protected by various sensors to avoid abnormal variations of the operating parameters. As a result, the only way to stress the klystron unit was to change its peak power output. So, it was decided to stress the unit by changing this single available parameter. The AE transducer was coupled to Location 3 (**Figure 8.1**) on the focusing coils of the klystron unit. Pulses with a frequency of 851 MHz were sent to the RF input connector of the klystron amplifier. The pulse repetition rate was set to 213 pps, with the pulses sent in pairs in rapid succession. The time interval between successive pairs of pulses was 4.69 ms. **Figure 8.3A** shows four separate signals on Zoom Trace A, which is the averaged acoustic AE signal from Channel 3. They corresponded with rises and falls of signals from the beam current sensor connected to Channel 2 of the oscilloscope. Channel 1 shows an approximately −50-mV output from the RF detector of two drive pulses fed into the input. The first narrow pulse, 2-μs wide, can be seen around the second vertical division. The second signal, 125-μs wide, is located immediately beyond the fifth vertical division. These two pulses are separated by an interval of 150 μs. **Figure 8.3B** shows the beam current for two pairs of pulses on Channel 2, separated

AE Transducer-R50

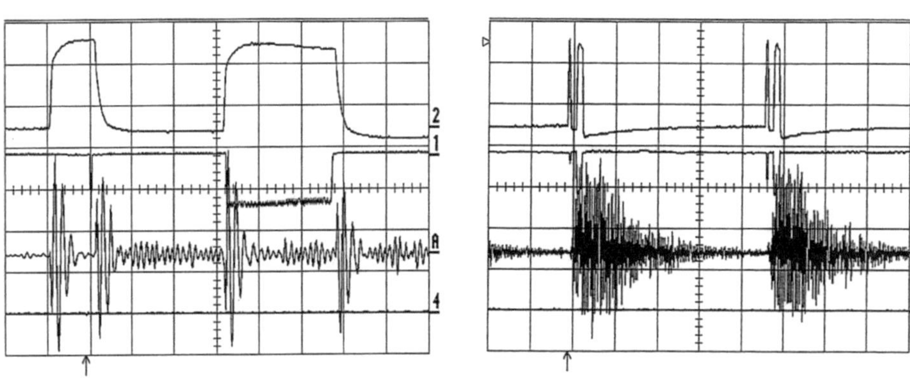

(a) (b)

Figure 8.3.
(a) Normal operation of klystron at frequency = 851 MHz. Ch 1: detected RF drive signals (−50 mV). These are klystron input signals. First-2 μs wide and second-125 μs wide separated by 150 μs. Ch 2: corresponding beam current pulses. Zoom Trace A: averaged AE signals from Ch 3 in sync with pulsing klystron. Ch 4: output of threshold amplifier (none). Time scale = 50 μs/div for Ch 1, 2, 4 and A. (b) Normal operation at 213 pulses per second. Ch 1: two pairs of input pulses are separated by 4.69 ms. Ch 2: beam current pulses. Zoom Trace A: averaged AE signals from Ch 3 in sync with pulsing klystron. Ch 4: output of threshold amplifier. Y-axis = 50 mV/div for Ch 1 and A for both **Figure 8.4A** *and* **B**. *Y-axis = 0.50 V/div for Ch 2 and = 10 V/div for Ch A for both* **Figure 8.4A** *and* **B**.

by 4.7 ms (4.7 divisions × 1 ms per division), which is the same time separation as for the RF drive signals on Channel 1.

Zoom Trace A, in **Figure 8.3A** and **B**, shows the average of the AE signals from Channel 3 that were generated by the klystron during pulsed operation. No AE signals appeared between the beam current pulses. Channel 4, in **Figures 8.4A** and **B**, shows the output of the threshold amplifier. The threshold amplifier produced no output signal during normal operation of the klystron.

After observing that Location 3 was giving strong AE signals, the transducer was relocated to Location 4 on the waveguide. In **Figure 8.4A**, Channel 3 shows the resulting AE signal. As in **Figure 8.3A** and **B**, Channel 1 shows the RF drive signal and Channel 2 shows the beam current sensor signal. **Figure 8.4A** shows the picture when the klystron was operated under normal conditions. The normal operating range for the beam current was 0.9–1 A. The peak power output was about 280 kW. Next, to stress the klystron unit, the output power was increased by increasing the beam current slowly to the level of 1.1 A. At this point, an abnormal current pulse suddenly appeared and the previously described threshold amplifier, on the output of the AE amplifier, responded by putting out a TTL high-level output on Channel 4, as shown in **Figure 8.4B**. Channel 2 showed an abnormal current pulse, with a significantly increased current amplitude, indicating a probable arc. This abnormal performance led to a decrease in the AE response (Channel 3), as expected. In **Figure 8.4A** and **B**, the vertical and horizontal sensitivities on Channels 1 and 2 are the same. However, the vertical sensitivity on Channel 3 was switched from 20 mV/div in **Figure 8.4A** to 50 mV/div in **Figure 8.4B**.

Even though the plots of AE signals in Channel 3 of **Figure 8.4A** and **B** are on different scales, a comparison between them shows that the AE signal amplitude in Channel 3 of **Figure 8.4B**, under abnormal operation, is much smaller than the one from **Figure 8.4A**. Therefore, these results experimentally show that the AE signal amplitudes

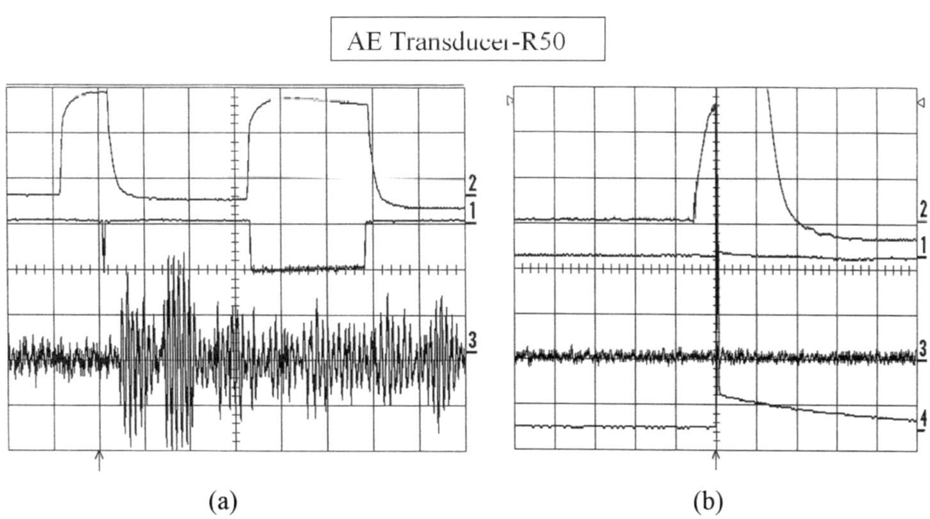

Figure 8.4.
(a) Normal klystron operation. Ch 1: one pair of RF drive pulses. Ch 2: normal klystron beam current pulses. Ch 3: Strong AE signal during normal operation. Vertical sensitivity = 20 mV/div. Ch 1: vertical sensitivity = 50 mV/div for both figures. Ch 2: vertical sensitivity = 0.50 V/div for both Figures. (b) Abnormal klystron operation. Ch 1: RF drive signals are not shown. Ch 2: abnormal klystron beam current pulse caused by stressing klystron. Ch 3: weak AE signal due to abnormal operation. Vertical sensitivity = 50 mV/div. Ch 4: vertical sensitivity = 10 V/div. Ch 4: output from threshold amplifier as a result of faulty klystron pulse. Time scale = 50 µs/div on all four channels for both A and B (after Joshi et al. [7]).

decreased during abnormal performance of the klystron unit and are similar to the results from the earlier tests on the magnetron and TWT amplifiers (TWTAs).

8.5. Experiments on a second klystron unit

With the encouraging results from the acoustic emission nondestructive testing technique for normal and abnormal operation of the first klystron unit, the next step was to repeat the experiments on a second klystron unit from a different location with different operating parameters. This second klystron was installed in a radar set that had a maximum operating range of 256 nmi. The radar system had two basic modes of operation, a long-range mode and anti-chaff mode. In these tests, the radar was operated in the long-range mode all the time. The transmitter of the system amplifies a 2-mW RF input drive signal from the receiver to a 9.8-kW average power level. This amplification was performed in two stages: through a driver and through the power klystron. After amplification, the RF energy from the klystron was routed to the antenna through either a coaxial cable or waveguide. The RF drive power rating to the klystron was 30 W. The klystron's filament required a voltage of 8 V. Its focusing coil required a current of 15 A. The frequency band used for these tests was the midband, which extended from 881.333 to 910.222 MHz. In addition to the fixed frequency mode, the radar also operates in the frequency agile mode. The fixed frequency used during this work was 902.999 MHz [5].

8.6. Experimental setup for the second klystron

The experimental equipment and arrangement was similar to that used with the first klystron unit (**Figure 8.2**). **Figure 8.5A** and **B** are photos of the equipment arranged near the klystron radar unit. The radar operated in the long pulse mode. The pulse repetition frequency (PRF) was set to 254.4 Hz, with a pulse repetition period of 3.93 ms and duty cycle of 0.0377. The peak RF power was 9.04 kW or 69.6 dBm. The short pulse width was 10 μs, and the long pulse width was 137.5 μs. The pulse separation was 160 μs. For these tests, the RF signal was fed to a dummy load at the end of the waveguide.

8.7. Experimental results from the second klystron

8.7.1 Collection of data during normal operation

Figure 8.6 identifies various locations at which the AE transducer was mounted to examine AE signal patterns generated by the RF pulse energy traveling through the waveguide. As in the previous tests, AE transducer R80 was used to pick up AE signals. As before, its output was connected through a pre-amplifier and post-amplifier, with the respective gains of 40 and 21 dB to Channel 3 of the oscilloscope. Connected to Channel 2 of the oscilloscope was a BNC T-connector with one side connected to the output of the klystron's current sensor (5A/V) and the other side to the previously described high pass filter. The output of the high-pass filter was connected to a detector, with the output of the detector going to the previously described low-pass filter. The output of the low-pass filter was connected to the previously described threshold amplifier with an output connected to Channel 4 of the oscilloscope. Zoom Trace C of the oscilloscope was used to show the averaged AE signal from Channel 3.

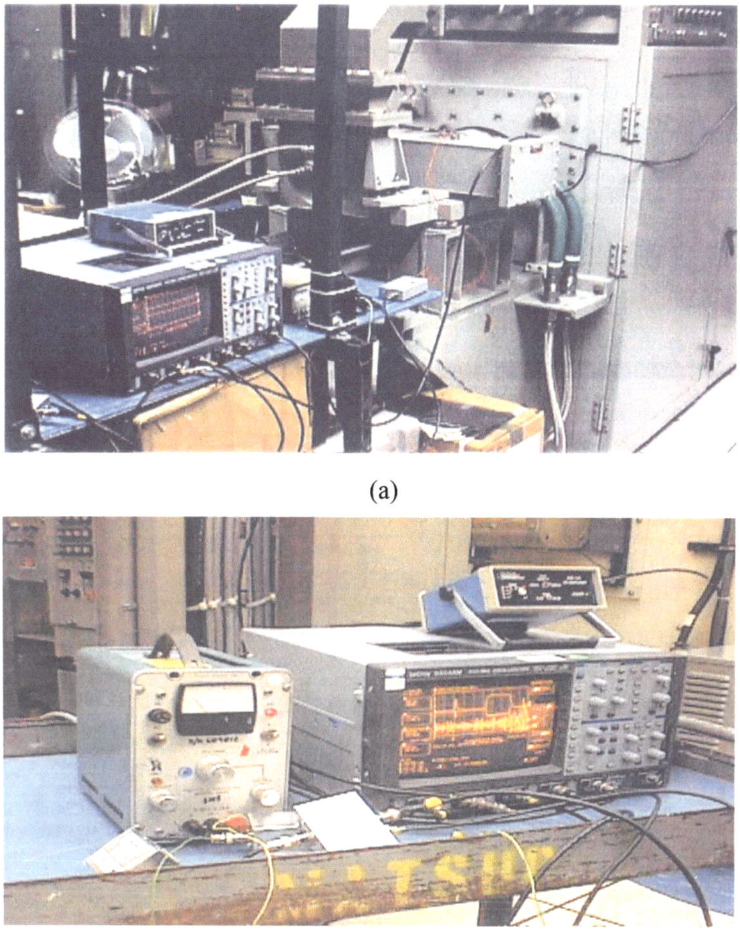

(a)

(b)

Figure 8.5.
(a) AE experimental setup (on left) with second klystron unit. The AE transducer is mounted on the low-pass filter box. (b) AE experimental setup for the second klystron unit. To the left of the LeCroy oscilloscope are the power supply and threshold amplifier. On top of the oscilloscope is the AE post amplifier.

8.7.2 Effect of different locations of AE sensor

To make sure that the equipment was not picking up extraneous signals from the air, the transducer was held in the air for a time, while the klystron current pulses were turned on. Channel C of the oscilloscope received no AE signal. The AE transducer was then coupled with a viscous couplant to the top of the low-pass filter at Location 4.

An AE signal in step with the current pulses was recorded on Channel C (averaged signal from Channel 3), as shown in **Figure 8.7A**. Channel 2 shows three pairs of klystron beam current pulses separated by about 3.93 ms. The horizontal time scale on Channel 2 was 1 ms/div. Acoustic emission activity of the burst type appeared on Channel C in step with the beam current pulses.

8.7.3 Data collection during stressed operation of klystron

For this test, the experimental setup was similar to the one used for the experiments with the first klystron unit. The combined gain of the AE pre-amplifier and

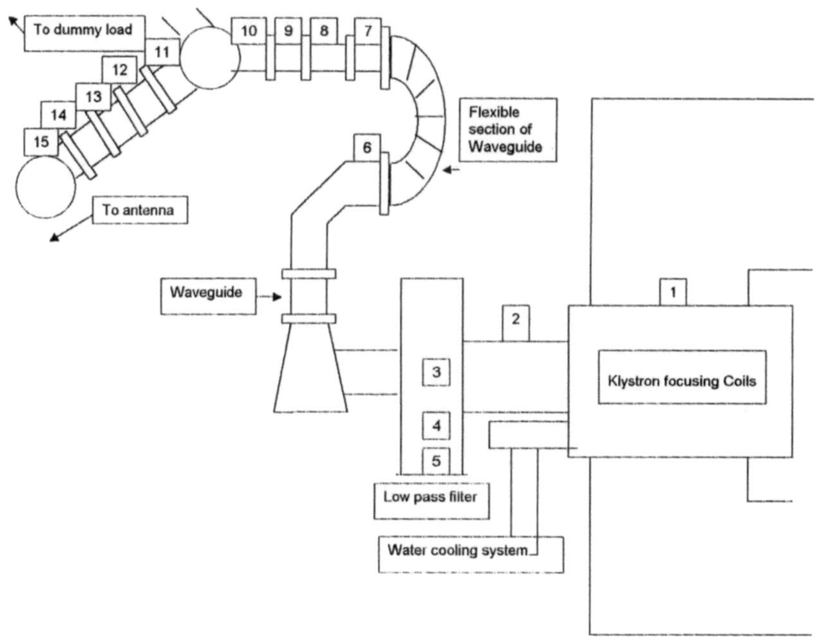

Figure 8.6.
Schematic diagram of AE transducer locations on the second klystron unit (after Joshi et al. [7]).

the post-amplifier was 61 dB, as before. The AE transducer was mounted on the coaxial cable at Location 2 (**Figure 8.6**).

At Location 8 on the in-line isolator in the waveguide, the AE activity recorded was the strongest for any of the locations, as seen in **Figure 8.8B**. The vertical sensitivity on Channel C was 50 mV/div in **Figure 8.8A**, while it was 0.50 V/div on Channel C of **Figure 8.8B**. During the collection of data from different locations on the waveguide, the gains on both AE amplifiers were held constant, so that the AE signal data from different locations could be compared. The AE transducer

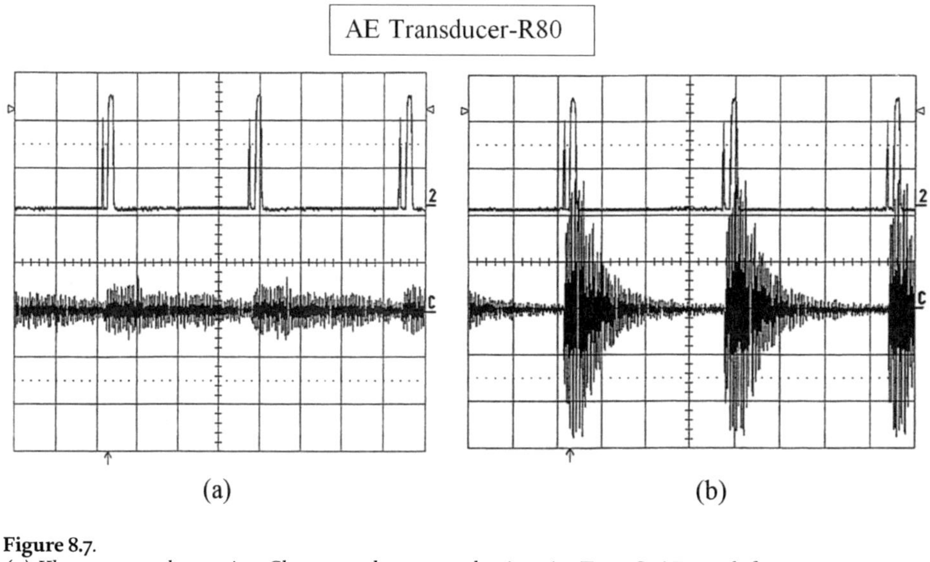

Figure 8.7.
(a) Klystron normal operation. Ch 2: normal current pulses in pairs. Trace C: AE signals from Location 4 on low-pass filter box. Trace C vertical sensitivity = 50 mV/div. (b) Normal operation. Ch 2: normal current pulses. Trace C: strong AE signals at Location 8 on isolator section of waveguide. Trace C vertical sensitivity = 0.50 V/div. Zoom Trace C showing the average AE signal from Ch 3 in both figures. Ch 2: vertical sensitivity = 2 V/div for both figures. Zoom Trace C: time scale: 1 ms/div for both figures (after Joshi et al. [7]).

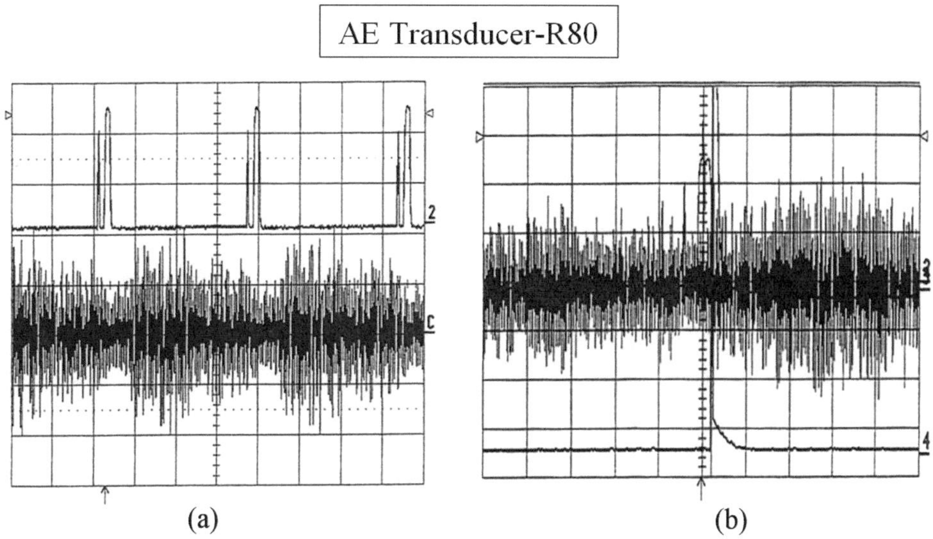

Figure 8.8.
(a) Normal klystron operation. Ch 2: normal current pulses—3 pairs. Ch 2: time scale = 1 ms/div. Vertical sensitivity =2 V/div on Ch 2 for both figures. Zoom Trace C: averaged AE signal from Ch 3. Trace C: time scale = 1 ms/div. Vertical sensitivity = 200 mV/div. (b) Abnormal operation. Ch 2: abnormal pulse-one pair. Ch 2: time scale = 0.5 ms/div. Ch 3: unaveraged AE signal. Ch 3: time scale = 0.5 ms/div. Vertical sensitivity = 50 mV/div. Ch 4: output of threshold amplifier was triggered by abnormal current pulse. Ch 4; time scale =0.5 ms/div and vertical sensitivity = 10 v/div (after Joshi et al. [7]).

was coupled carefully at each new location to make sure that no air bubbles were trapped in the couplant layer.

The beam current was slightly less than 1 A. AE activity was recorded during normal operation as shown in **Figure 8.8A**. Channel 2 shows the beam current pulses, and Channel C shows the averaged AE signal. Next, the high voltage was gradually increased beyond the normal value of 40 kV. The current showed a slight increase beyond 1 A, when a TTL high-level output from the threshold amplifier was recorded on Channel 4, as seen in **Figure 8.8B**. Channel 3 shows the recorded AE signal from abnormal operation before averaging. An averaging operation was then performed to remove noise from the signal. The vertical sensitivity was set at 200 mV/div on Channel C in **Figure 8.8A** (normal operation), while it was set at 50 mV/div on Channel 3 in **Figure 8.8B** (abnormal operation). The comparison of strengths of two AE signals shows that during abnormal operation, the strength of AE signal decreased, provided all other parameters were kept unchanged.

Figure 8.9A shows an abnormal beam current pulse from Channel 2. The horizontal time scale was expanded to 0.5 from 1 ms/div. Signals from other channels were not displayed. In **Figure 8.9B**, the time scale was further expanded to 50 μs/div on Zoom Trace A of Channel 2 to see the fine structure of the abnormal pulse. In this pair of consecutive pulses, the first pulse is 10 μs wide, as seen just before the first vertical line. The second pulse started just before the fourth vertical line. The separation between these consecutive pulses was 160 μs, as expected. The second pulse, of 137.5 μs width, developed a dip in its structure, as can be seen on the fifth vertical line. This abnormal structure in the second beam current pulse caused the threshold amplifier to generate a high-level TTL output. Only the TTL output from Channel 4 is shown in **Figure 8.10A**, with a horizontal scale of 0.5 ms/div. **Figure 8.10B** shows the same TTL output with an expanded horizontal scale of 4 μs/div on Zoom Trace A of Channel 4 from **Figure 8.10A**. A normal or good pulse will not trigger the threshold amplifier, which was connected to Channel 4. It takes its input

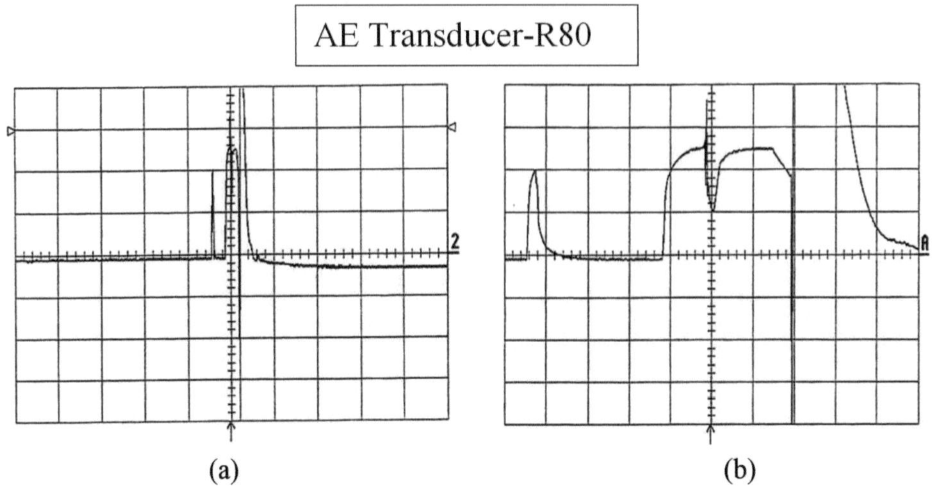

Figure 8.9.
(a) Abnormal current pulse. One pair on Ch 2. Ch 2: time scale = 0.5 ms/div. Vertical sensitivity = 2 V/div. (b) Expanded view is on Zoom Trace A of Ch 2. Trace A: time scale =50 μs/div. Vertical sensitivity = 2 V/div. Width of first pulse in the pair =10 μs and that of second pulse = 137.5 μs. Separation of pulses in the pair = 160 μs. The second pulse is distorted (after Joshi et al. [7]).

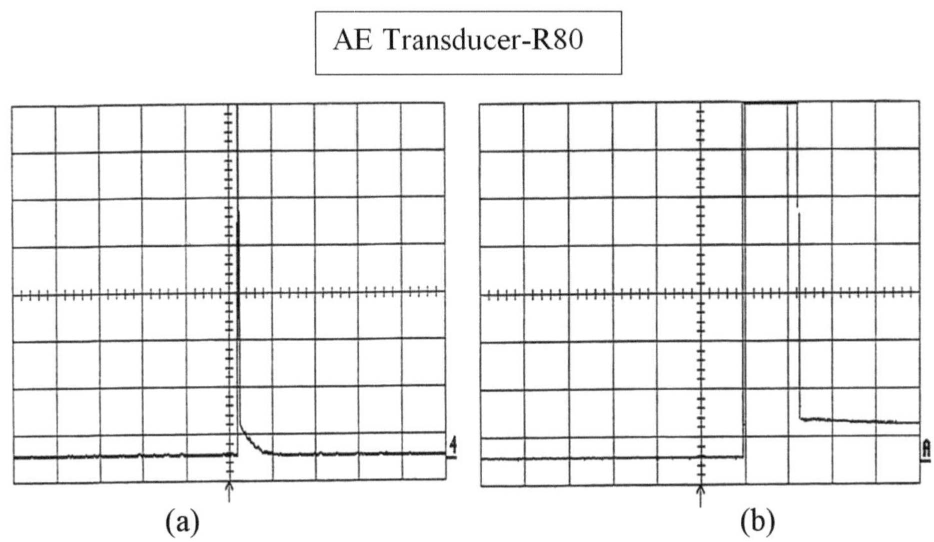

Figure 8.10.
(a) Output from threshold. Amplifier triggered by abnormal pulse. Ch 4: time scale = 0.5 ms/div. Vertical sensitivity = 0.5 V/div. (b) Expanded view on Zoom Trace A of Ch 4. Trace A: time scale = 5 μs/div. Vertical sensitivity = 0.5 V/div (after Joshi et al. [6]).

from Channel 2 (of the klystron beam current pulses) through the high-pass filter, detector, and low-pass filter. An abnormal pulse produced by stressing the klystron tube beyond its normal operating range has high-frequency components. The high-pass filter (beyond 1.6 MHz) stops regular pulses but allows higher-frequency components of abnormal pulses to go through. The detector then detects these higher frequency components above 1.6 MHz and passes its output to the low-pass filter (below 15 MHz), which cleans up the signal. The output of the low-pass filter was then passed to the threshold amplifier, which responded by generating a high-level TTL output. Further details on the electronic circuits for detection of abnormal pulses can be obtained from Joshi et al. [6].

8.8. Conclusions from magnetron, TWTA, and klystron tests

The advanced nondestructive technique of acoustic emission sensing was applied for the first time in this research project to the performance monitoring of four high-power microwave tubes: a magnetron, a traveling wave tube amplifier, and two klystron amplifiers. In all four cases, strong "burst-type" acoustic emissions were generated during pulsed operation of the tubes. The TWTA, when operated in the CW mode, generated only sporadic signals that could not be associated with any definite source mechanisms. Irregular pulses generated by the magnetron were clearly distinguished from regular pulses using different features of their associated AE patterns. Cold cathode operation of a TWTA-generated AE activity bursts in quick succession. Positioning the AE transducer at different locations on the waveguides of the klystron units produced remarkably different AE patterns. In all the experiments with different high-power microwave tubes, abnormal operation of the tubes showed a decrease in the strength of their AE emissions when compared with AE emissions from tube operating under normal conditions. These investigations, using only one AE sensor to monitor in situ performance of high-power microwave tubes, have certainly proven AE emission sensing to be a great advancement over the currently available data collection systems that use as many as 11 electrical sensors to measure different currents and voltages of tubes.

Acknowledgements

Figures 8.2, 8.4, 8.5, 8.7–8.10 are printed with the permission of American Society for Nondestructive Testing Inc.

References

[1] Ewell GW. Radar Transmitters. McGraw Hill Book Company; 1981. p. 56

[2] Roach B. Klystrons. In: Koryu Ishii T, editor. Handbook of Microwave Technology. Vol. 2. Academic Press; 1995. p. 2

[3] Cole HW. Understanding Radar. 2nd ed. Cambridge, MA: Blackwell Scientific Publications Inc. p. 105

[4] Coleman JT. Microwave Devices. Reston, Virginia: Reston Publishing Company; 1982. p. 122

[5] Technical Manual for Radar Sets AN//SPS-49 (V)5 and (V)7. Vol. 1. Naval Sea Systems Command; 1989 and 1993

[6] Joshi NR, Brock DW, Russell SD, Lasher ME, Kasa SD. Built-in Test for High Power Microwave Tubes Using Acoustic Emission Technology. SSC Technical Note 1816. San Diego, CA: SSC San Diego; September 2001

[7] Joshi NR, Russell SD, Ramirez AD, Brock DW. Performance monitoring of high power microwave radar klystron units with acoustic emission technology. Materials Evaluation. 2007;**65**:1048-1053

Chapter 9

Interfacing and Integrating Acoustic Emission Data (Part I)

9.1. Integration of AE data with a computer

High-power microwave tube systems used in radar and communication systems are expensive. Expenses related to these systems are primary cost drivers, with annual costs reaching about $100 million spent on replacing malfunctioning high-power microwave tubes used in various radar installations. In many cases, it is estimated that more than 25% of perfectly operating tubes are inadvertently replaced because in situ monitoring equipment is unavailable to diagnose specific problems with these systems. Sometimes, high-power microwave vacuum tubes used in field operational systems are equipped with only limited online capability for assessing the condition of system components. With little or no built-in prognostic capability, the users have no advance warning of a developing failure situation.

Some companies design microprocessor-based systems with as many as 11 sensors to monitor tube performance, provide tube protection, and record a comprehensive tube failure history. As described before, a major limitation with these systems is the small amount of time available during the tube interpulse period for data buffering and fault analysis. These monitoring systems work well, if the microwave tube is operated at rates of 200 or less pulses per second. However, these tubes are typically operated at up to 1000 pps, with a 1-μs pulse. Increasing the data acquisition speed will, in some cases, make the situation worse, since it will increase the amount of data that must be transferred and analyzed during the small time interval available. Data from acoustic emission (AE) systems is instantaneous, however, and can be monitored with an advanced digital oscilloscope.

This study successfully demonstrated that the new advanced AE nondestructive testing technique has the potential for in situ monitoring of normal and abnormal performance of high-power microwave tubes such as magnetrons, traveling wave tubes (TWTs), and klystrons. However, in many cases, radar installations using microwave tubes are monitored simultaneously with other mechanical and electrical systems, which interface with graphical user interface (GUI)-driven software on personal computers (PCs). Recently, new techniques using virtual sensors have been developed to facilitate certain decision-making processes in cases of malfunctioning equipment. Consequently, it is starting to become imperative to interface AE equipment with currently available GUI software. This interfacing of AE data requires the development of the appropriate hardware and software, including the development of electronic systems for the conditioning of AE data, which is necessary before it can be fed to the computer through data acquisition interfacing hardware. This chapter describes the efforts made in this direction.

9.2. Data acquisition unit (DAU)

An intermediate unit, capable of taking raw data from the current sensors and AE sensors attached to the microwave tubes and transferring it to the host computer for further analysis and integration, was required. The Opto 22 interface unit was already used to obtain data from mechanical systems such as pumps and turbines on commercial and naval ships. It was decided therefore to use this same interface unit, called an Opto 22 SNAP B3000 brain, in this effort. The B3000 unit can be used with either an Opto 22 controller or a host computer. The unit with the host computer was chosen. The B3000 communicates serially with a host computer over RS-485 twisted wire pairs. It supports the advanced Mistic (popular bulletin board system) protocol and the industry standard Optomux protocol, with both protocols able to support high-speed communication (115 kBd). This second protocol, the Optomux, was used.

An AC37 adapter card was installed in one of the industry standard architecture (ISA) slots inside the host computer to facilitate an RS-485 two- or four-wire communication link between the computer and Opto 22 remote input/output (I/O) bus. The remote bus is a standard RS-485 half-duplex, two-wire serial communications link. The AC37 is designed to appear as a standard IBM serial "COM" port to user software. The AC37 has two operating modes, Mode 0 and Mode 1. While in Mode 0, the AC37 will operate very much like a standard serial COM port. Mode 0 was selected here by removing the Opto jumper on the card. **Figure 9.1** shows the schematic arrangement of the B3000 Optomux System Architecture.

The Opto 22 SNAP B3000 brain was connected to a SNAP B series I/O rack capable of holding eight SNAP modules, as shown in **Figure 9.1B**. In this effort, only one analog voltage input module, with a – 10- to +10-VDC range (SNAP-AIV with two channels) was used in the "Module 0" position, next to the SNAP B3000 brain. The analog input module is transformer isolated as well as optically isolated from all the other modules and from the SNAP B3000 brain. Transformer isolation prevents ground loop currents from flowing between field devices and causing noise that produces erroneous readings. Optical isolation provides 4000 V of transient protection to the sensitive control electronics from real-world industrial field signals. However, the two channels of the analog input module use a common reference and are not isolated from each other. The SNAP analog input module has an on-board microprocessor to provide module-level intelligence.

The SNAP-PS5 power supply takes 120 VAC input from the power line and provides 5-VDC power for loads up to 4 A. It supplied power to the SNAP B3000 brain and SNAP I/O module through the connector strip mounted on the rack. The power supply module did not sit on the rack but was instead simply placed next to it.

The connectors and jumpers on the top of the SNAP B3000 brain were selected to adjust the baud rate to 19.2 kBd and the analog address to 02 and to set the Optomux communication protocol. Each rack had an address of the form 0, 1, 2, etc. Since we were using rack 1, its address was set to 0. I/O on the SNAP B3000 brain was divided into four addresses (two digital I/O and two analog I/O). The digital addresses are base+0 and base+1. The analog addresses are base+2 and base+3, so that the analog module snapped onto the rack has the total address of 0+2.

The SNAP B3000 brain is connected to an adapter card inside the computer. Details are provided in the Optomux Protocol User's Guide [1].

Optomux is a protocol used by a family of intelligent digital and analog I/O units that operate as slave devices to a host computer. The Optomux protocol can be used to communicate at eight different baud rates with the rate 19.2 kBd chosen because it is suitable for other advanced software packages, in addition to the software that came with the Opto 22 unit. To read data from the computer, the Optoscan (OptoDriver Toolkit) software was provided. The user needs to enter the baud rate (19.2 kBd), COM port (3), and the address (02) to view the table showing the voltage values (between 0 and 10 VDC) from the two channels of the analog input module. Electronic circuits were interposed between the sensor signals (from both the current sensor and AE sensor) and the analog input module on the rack. The output voltages of these interface circuits ranged from 0 to 10 V.

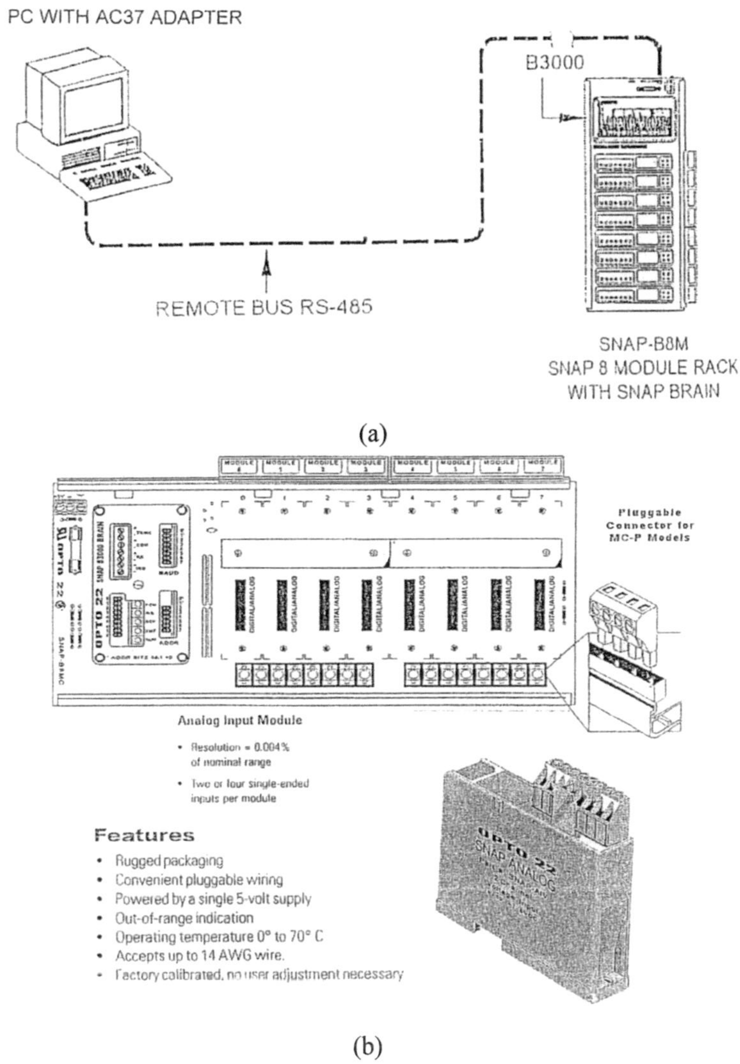

Figure 9.1.
(a) Schematic drawing of the B3000 Optomux System Architecture. (b) SNAP B series I/O rack with SNAP B3000 BRAIN on left (top) and OPTO 22 analog input module (bottom).

9.3. Interfacing electronic circuits

9.3.1 Interface for faulty current pulses

The aim of the experiment was to distinguish the faulty current pulses from the normal current pulses and their corresponding weak AE signals from the strong AE signals that occur during normal operation of the high-power microwave tubes. To perform this task, the plan was to develop circuitry to generate high-level transistor-transistor logic (TTL) output pulses corresponding to bad current pulses with a low-level TTL output for good current pulses. **Figure 9.2** shows a block diagram of electronic circuits that could achieve this aim. The circuits are identified with the numbers included in square brackets [1C] to [5C], where the suffix "C" stands for "Current." The voltage pulse input to the magnetron came from an Interstate Electronics Corporation Model P25 Pulse Generator. This pulse generator was triggered by an external trigger circuit. Current pulses were generated at the rate of 60 pps, synchronized with the power line.

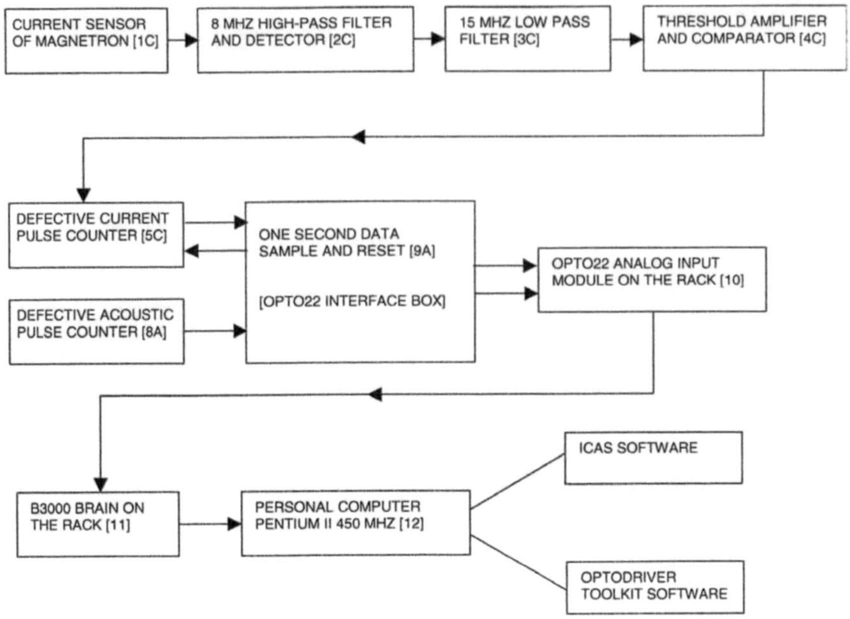

Figure 9.2.
Block diagram of electronic circuits between current sensor and OPTO22 interface box for detecting and counting defective current pulses.

In this work, the magnetron unit was used as the source of high-power microwave pulses. In future tests, interfacing circuits could be modified for use with different microwave tubes. The normal cathode current pulses in the magnetron have very little spectral content above a frequency of about 8 MHz, while bad pulses have a much greater spectral content above 8 MHz. Thus, an 8-MHz high-pass filter (Block [2C] as shown in **Figure 9.3**) may produce a much greater output for bad pulses compared to that for good pulses. The signal was then filtered by a 15-MHz low-pass filter (Block [3C] as shown in **Figure 9.4**), which cleaned up the spectral content of the pulses and smoothed them, but with the output, after going through the 15-MHz filter, still being relatively small for good pulses and much larger for bad pulses. The output from the low-pass filter was then fed into the threshold amplifier (Block [4C] as shown in **Figure 9.5**) that had a comparator with an adjustable threshold voltage that could be set to a level where it would not respond to good pulses but would generate a 5-V TTL-compatible output pulse when a bad current pulse occurred.

Next, electronic circuits were needed to count the faulty current pulse signals from the threshold amplifier. Two methods were developed to do the counting: a digital method for usage when reset pulses were available and an analog method for usage in the absence of reset pulses. Electronic circuits for these two counting methods are shown in the circuit diagram, “Defective Current Pulse Counter” in (Block [5C] as shown in **Figure 9.6**), with the upper part of the drawing showing

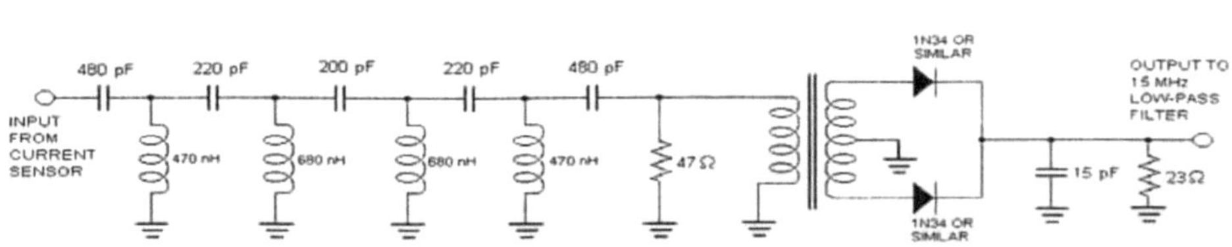

Figure 9.3.
Circuit of 8-MHz high-pass filter and detector [2C].

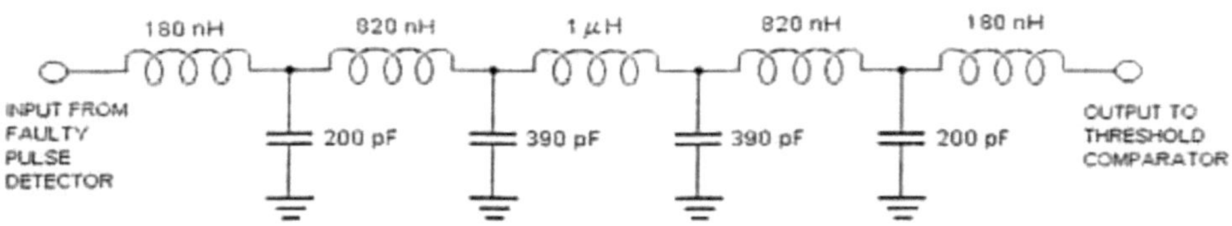

Figure 9.4.
Circuit of low-pass filter—cut-off, 15 MHz [3C].

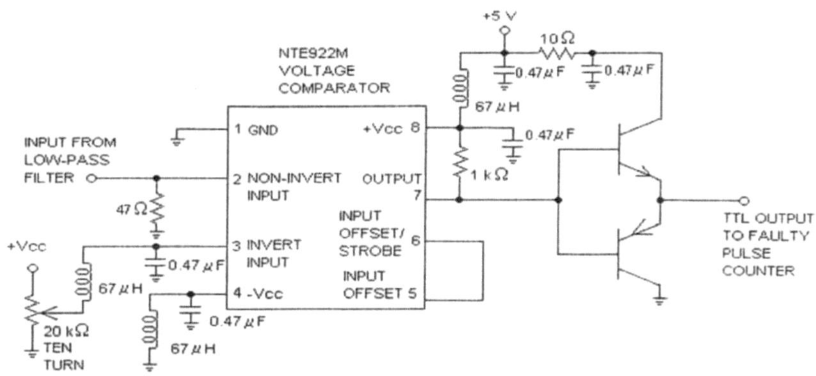

Figure 9.5.
Circuit of threshold amplifier and comparator [4C].

the digital section and the lower part, the analog section. The digital section used two serially connected TTL count-by-ten circuits, giving the counter the ability to count up to 100 faulty pulses between reset pulses. The output of this two-stage TTL counter was fed into a digital-to-analog converter, since the host computer system monitoring software was set up to use only analog inputs. After each reset pulse, these TTL counters would begin counting pulses again from zero.

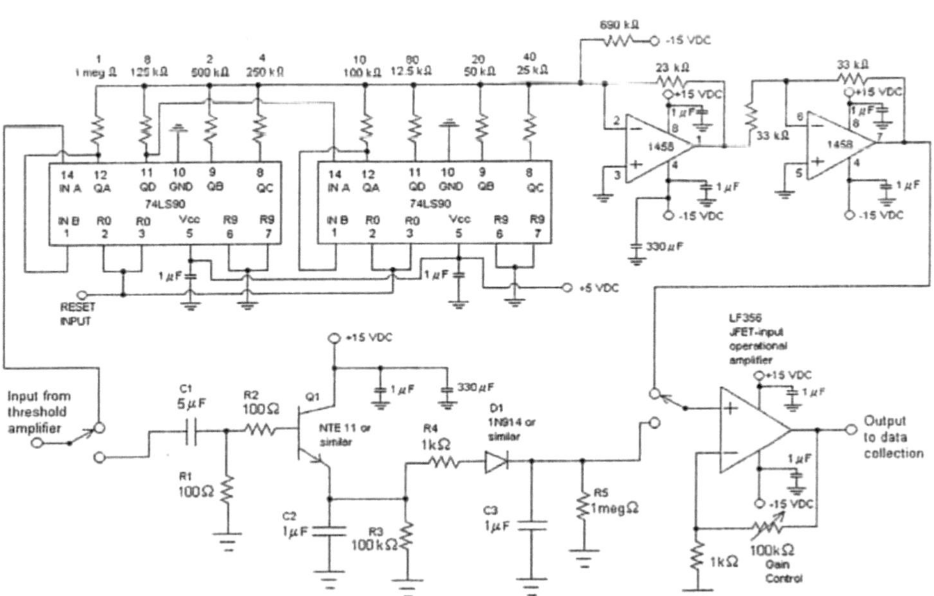

Figure 9.6.
Circuit of defective current pulse counter [5C].

The analog pulse counter circuit is shown in the lower part of the diagram. One problem with counting pulses from the threshold amplifier is that the lengths of the pulses vary from 1 to 3 μs. This problem was solved by sending the input pulses to the base of a high-gain, high-current transistor, Q1. The transistor fully charges a 1-μF capacitor, C2, connected to its emitter, in a time period less than 1 ms, so that if the pulses are longer than 1 μs, there will be no effect on the charge on capacitor, C2. This capacitor then discharges through resistors R3 and R4 over a time of approximately 1 ms. During this discharge period, most of the current passes through a 1-kΩ resistor, R4, and a diode, D1, into the second 1-μF capacitor, C3. This capacitor serves as a charge accumulator and analog counter of the number of faulty pulses.

Counter capacitor C3 is connected to the high-impedance junction field-effect transistor (JFET) input of an LF356 operational amplifier that measures and amplifies its voltage while not discharging the capacitor. A method must be provided to discharge capacitor C3. Otherwise, the voltage on capacitor C3 would quickly reach a maximum value and stop accumulating charge and stop counting. The discharge method is used to continuously discharge the capacitor through a 1-MΩ resistor, R5, which results in a 1-s time constant for the discharge time. This results in an exponentially decaying voltage leading to a time-weighted averaging method, which may be unsuitable for some applications.

9.3.2 Interface for corresponding faulty acoustic pulses

Figure 9.7 is a block diagram of the electronic circuits designed to detect and count faulty acoustic pulses. These circuits are identified with the numbers [1A]–[9A], where the suffix "A" stands for "Acoustic." The AE signals generated during the operation of the pulsed magnetron were picked up by the AE transducer, shown as Block [1A]. It was experimentally determined that the total acoustic pulse energy of faulty pulses was significantly less than that of good pulses. We determined the total energy of acoustic pulses coming from the acoustic energy amplifier by passing the output of the post-amplifier of Block [A3] into an acoustic energy integrator, shown as Block [4A] in **Figure 9.8**, which rectified the acoustic signal, and then used the rectified output to charge a capacitor. The output of the capacitor was then sent through a 10-kHz low-pass filter (Block [5A] as shown in **Figure 9.9**), to clean up the output pulses. The output from the low-pass filter was then fed into a circuit containing a time delay, amplifier, and sample-and-hold (Block [6A] as shown in **Figure 9.10**). As shown in this circuit diagram, the operational amplifier U3 amplified the voltage of the pulses by a factor of 33, while the timer, integrated circuit U1 generated a time delay that allowed the acoustic pulse integrator time to fully integrate the energy of the acoustic pulses. Its delay time could be adjusted from 0 to 3 ms and would typically be set for about 1 ms. At the end of the delay period, U2 would send a 100-μs pulse to the sample-and-hold integrated circuit, U4, which would sample the output from amplifier U3. Also shown in the diagram is a 50-Ω output driver circuit, consisting of operational amplifier U5 and the MJE305 and TIP42C transistors, which boosted the current of the output of U4. The output of the circuit was then sent through a second 10-kHz low-pass filter (Block [7A] as shown in **Figure 9.11**), which removed unwanted voltage spikes from its output.

The next step was to count the faulty acoustic pulse signals, as shown in Block [8A] in **Figure 9.7**. Two methods were developed to do the counting that were similar to those used for the faulty current pulses. A digital method was developed for usage when reset pulses were available, and an analog method for usage in the absence of reset pulses. These two counting methods are represented in **Figure 9.12**. The upper part of the drawing shows the digital section and the lower part, the analog section. A selector switch directs pulses to either section. The digital method

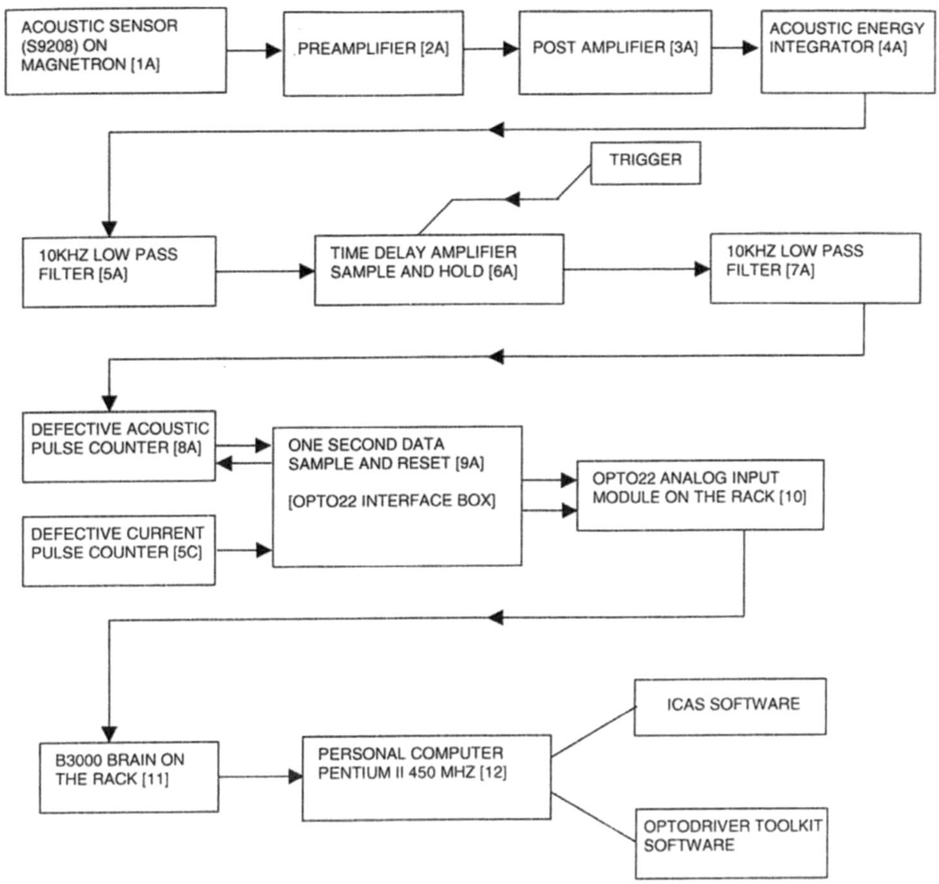

Figure 9.7.
Block diagram of electronic circuits between AE sensor and Opto 22 interface for detecting and counting defective acoustic pulses.

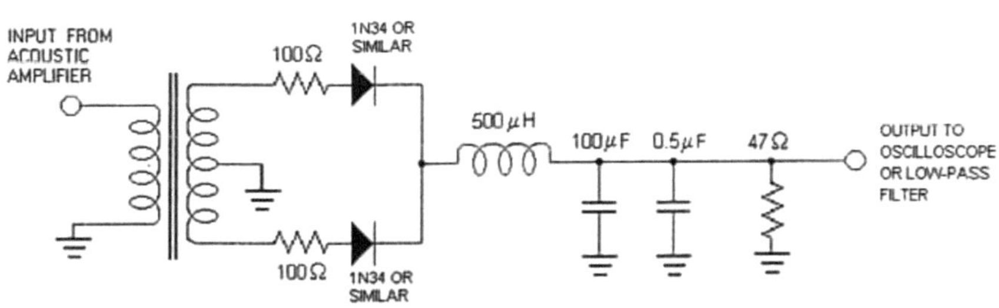

Figure 9.8.
Circuit of acoustic pulse energy integrator [4A].

used two TTL count-by-ten circuits connected in series, forming a counter with the ability to count up to 100 faulty pulses between reset pulses. The outputs of the TTL counters were fed into a digital-to-analog converter, since the host computer system monitoring software was set up to use only analog inputs. After each reset pulse, the TTL counters would begin counting pulses again from zero.

The analog pulse counter circuit is shown in the lower part of the diagram. The input from the faulty pulse detector circuit is fed through capacitor C1. During normal operation of the magnetron, the input voltage coming from sample-and-hold integrated circuit U4 of Block [6A] has a constant positive value; however, when

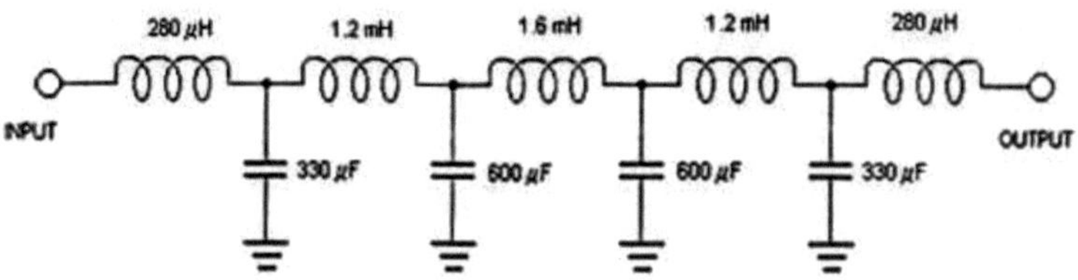

Figure 9.9.
Circuit of low-pass filter—10-kHz cut-off [5A].

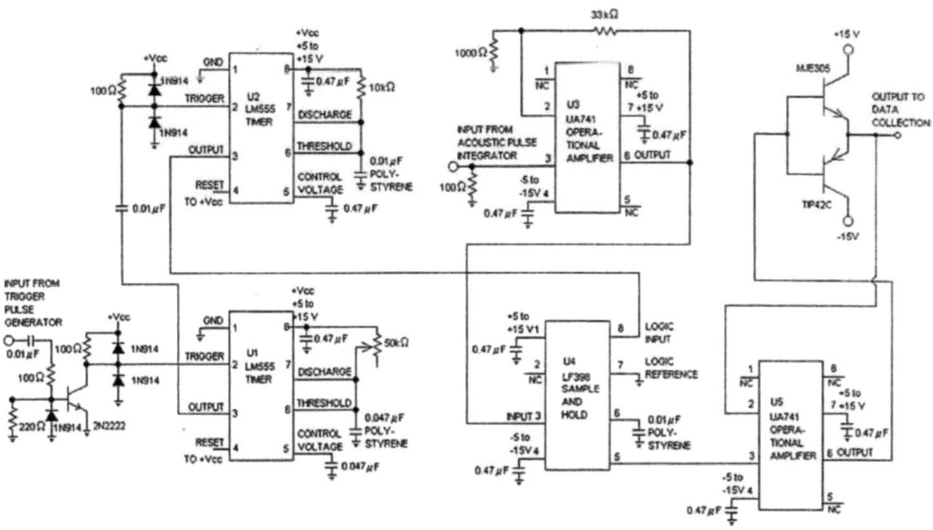

Figure 9.10.
Circuit of time delay, amplifier, and sample-and-hold [6A].

faulty pulses occur, this positive voltage is reduced. The amount of voltage reduction varies from one faulty pulse to the next, so that the circuit has to compensate for these variations. It does this by clamping the output of operational amplifier U1 to the range of −0.15 to +5.0 V. Capacitor C1 introduces DC blockage, so the inverting input of U1 will receive a negative input pulse when a faulty pulse occurs. Potentiometer R3 provides an adjustable positive bias on the inverting input of U1. During good pulses, the output of U1 will remain at a constant negative voltage, with diode D1 limiting this voltage to −0.15 V. When a faulty pulse occurs, the negative voltage from C1 will overcome the positive bias from R3, at the inverting input of U1, so that the output of U1 will become positive. The gain of U1 is adjusted to a high value, so that every faulty pulse will drive its output to the maximum positive limiting value of +5.0 V. When the switch in the circuit is set to the lower position for analog counting, the output current of U1 will flow through a 1-MΩ resistor, R7, and diode D2, into a 1-μF capacitor C2. This capacitor serves as a charge accumulator and analog counter of the number of faulty pulses. C2 is connected to the high-impedance JFET input of operational amplifier U2, which measures and amplifies its voltage while not discharging the capacitor. A method must be provided to discharge capacitor C2. Otherwise, the voltage on capacitor C2 would quickly reach a maximum value and would consequently stop accumulating charge and stop counting. The discharge method used is to continuously discharge the capacitor through a 1-MΩ resistor, R8, which results in a 1-s time constant for the discharge. This results in an exponentially decaying signal, leading to a time-weighted average, which may be unsuitable for some applications.

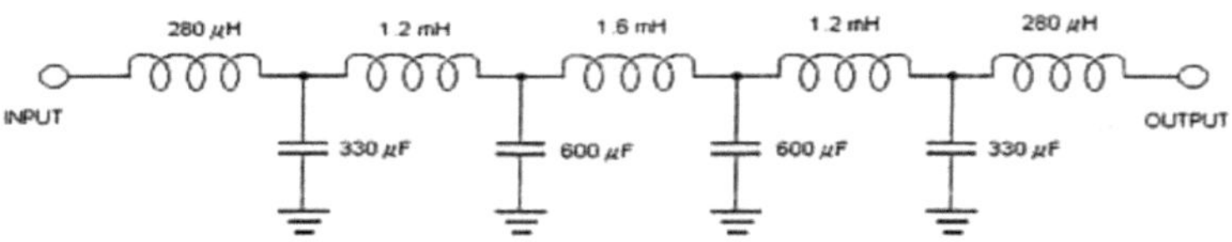

Figure 9.11.
Circuit of low-pass filter—10-kHz cut-off [7A].

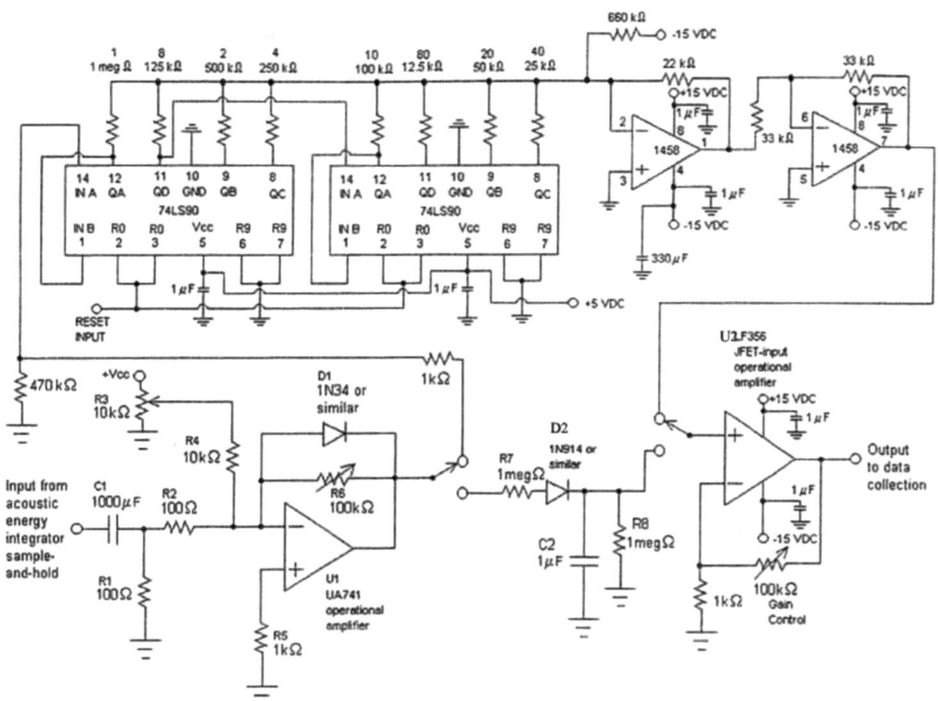

Figure 9.12.
Defective acoustic pulse counter [8A].

The Integrated Condition Assessment System (ICAS) software installed on the host computer is different from the software that came with the Opto 22 unit, which was discussed in Section 2. The ICAS software checks data once every second and no reset pulse was available. Consequently, it was necessary to develop a circuit with a timer, a sample-and-hold, and an amplifier to provide once-per-second data sampling from the faulty current and acoustic pulse counters and to provide reset pulses for the counters. This circuit called "One Second Data Sample and Reset" is shown in **Figure 9.13**. It is shown as Block [9A] in **Figures 9.2** and **9.7** and is common to both current sensor and acoustic sensor data. It is also called the Opto 22 interface circuit. The timer section consists of four LM555 timer integrated circuits. The first timer, U1, provides a 1-s period of time for counting the pulses. Its time period is adjustable with a 1-MΩ potentiometer. The second timer, U2, provides a 100-μs output pulse, which is fed to the inputs of two sample-and-hold circuits, U5 and U6. U5 samples and holds the output of the faulty current pulse counter, while U6 samples and holds the output of the faulty acoustic pulse counter. Timer U3 provides a waiting period of 25 μs to ensure that U5 and U6 have passed beyond their aperture times, before resetting the counters. This prevents any changes on the input voltages, which could affect the output voltages. Next, timer U4 provides a 25-μs voltage pulse to the reset inputs of the faulty current and acoustic pulse counters. Timer U1

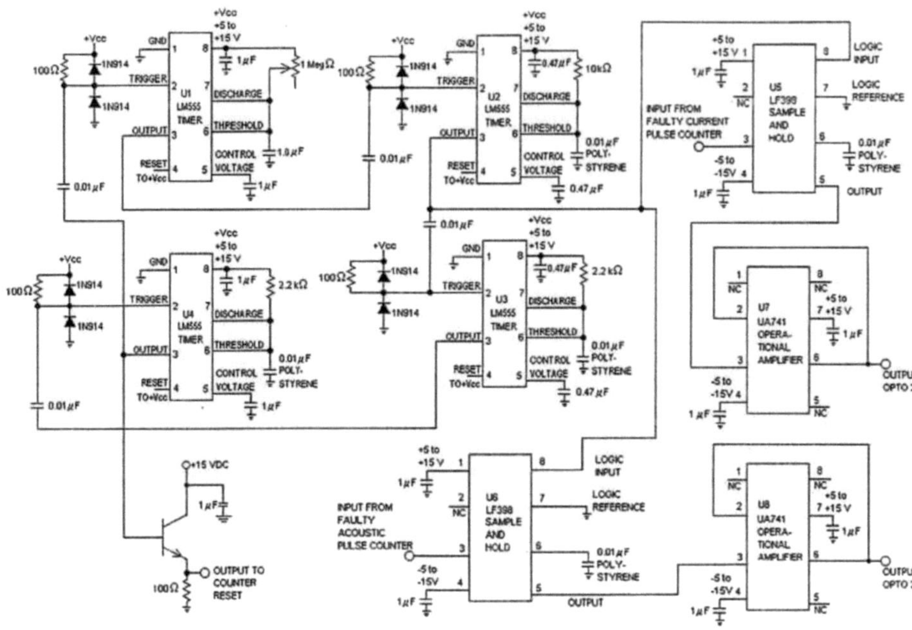

Figure 9.13.
Circuit of one second data sample and reset [9A] Opto 22 and ICAS interfacing circuit.

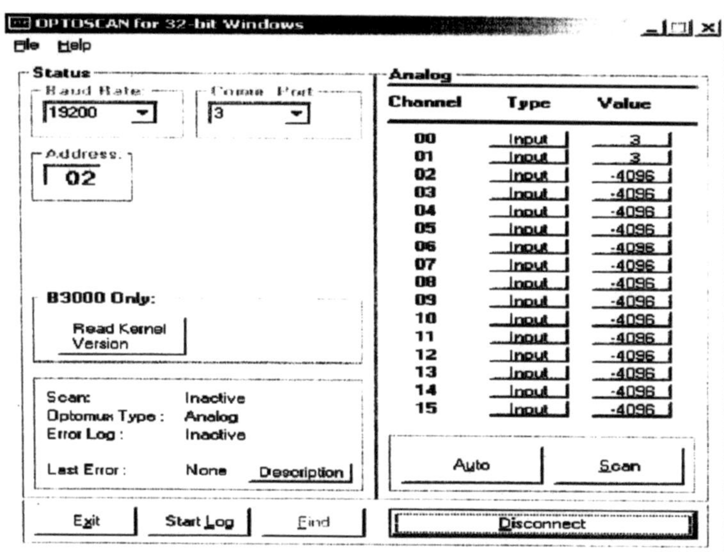

Figure 9.14.
Simple OptoScan Window (by Opto 22) to see values of the sensors 00 and 01.

then starts a new cycle, providing a new interval for counting the faulty current and acoustic pulses. The outputs of U5 and U6 are fed to operational amplifiers U7 and U8, respectively, which serve as buffers. The outputs of amplifiers U7 and U8 are fed to the Opto 22 Analog Input Module on the rack.

9.3.3 Opto 22 software

The OptoDriver Toolkit of the Opto 22 equipment can be used for direct communication from the host computer to the SNAP B3000 brain. The Toolkit included a 32-bit Windows driver. With the OptoScan (for 32-bit Microsoft® Windows Version A4.0b) window, the operator can see the values of the two sensor outputs, for current and acoustic emission, listed in the table of sensors. However, this software is extremely limited and could not be used for other computations that may be needed. **Figure 9.14** shows a view of the simple Opto 22 OptoScan window showing the values of sensors 00 and 01. The window also shows the baud rate as 19,200, the COM port as 3, and the address as 02.

9.4. Experimental procedure

9.4.1 Checking the Opto 22 Hardware and its OptoDriver Toolkit Software

The first step in the procedure was to check the functioning of the Opto 22 Analog Module and its accompanying Opto 22 software. The module was installed on the rack next to the Opto 22 SNAP B3000 brain, which in turn was connected to the AC37 adapter card installed in the host PC. Two DC voltages were fed into the two DC power inputs of the analog module from two separate HP6216A power supply units. After all the connections were completed, the units were turned on. The software then accurately acquired the simulated sensor voltages, as can be seen from the first two channels, 00 and 01, each showing values of 3 V in the rightmost column of **Figure 9.14**. These results confirmed that the Opto 22 system was working. The next step was to check all the electronic circuits between the two sensors and analog input module of the Opto 22 unit.

9.4.2 Checking of various electronic circuits

The outputs of various electronic circuits were checked, one at a time to ensure they were functioning properly before connecting them to the other circuits, as shown in **Figures 9.2** and **9.7**. The inputs and outputs of the circuits tested were connected to the various channels of a LeCroy 9354 AM 500-MHz digital oscilloscope. This oscilloscope has four regular channels (1, 2, 3, and 4) and four zoom traces or channels (A, B, C, and D). The sampling rate is 2 GS/s, with a storage capacity of one million points if a single channel is used, and 500 MS/s, with 250-KS storage capacity per channel, if all four channels are used. The device under test, a pulsed magnetron, was then turned on. An AE transducer (S9208) was coupled to

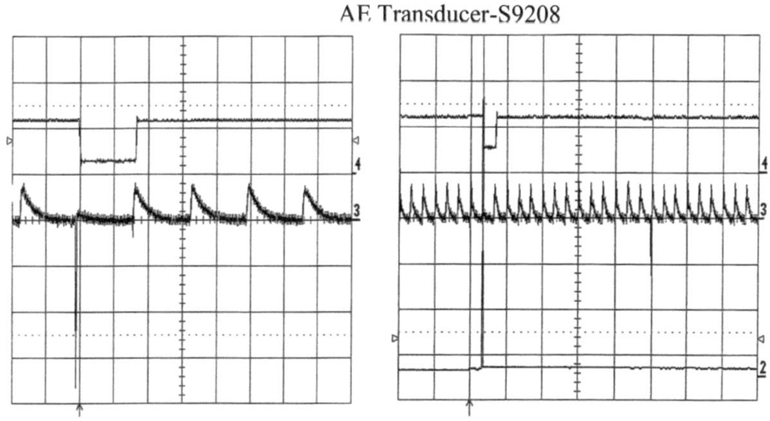

Figure 9.15.
(A) Defective acoustic energy pulse on Ch 3 at the vertical division 2. Ch 4: corresponding TTL output. Ch 3 and 4 time scale = 10 ms/div. Ch 3: vertical sensitivity = 50 mV/div. Ch 4: vertical sensitivity = 1 V/div. (B) Defective current pulse on Ch 2. Ch 3: defective acoustic energy pulse. Ch 4: corresponding TTL output. Ch 2, 3, and 4: time scale = 50 ms/div. Ch 3: vertical sensitivity = 50 mV/div. Ch 2 and 4: vertical sensitivity = 1 V/div.

the magnetron's anode cylinder with a couplant and held in position with a clamping mechanism. Any AE signal picked up by the transducer during the pulsed operation of the magnetron was fed to the pre-amplifier. The useful frequency range of the pre-amplifier was 100 kHz to 1 MHz. The signal was then amplified by the post-amplifier and then fed to the acoustic energy integrator circuit, shown in **Figure 9.9**. Earlier research work showed that only very aged high-power microwave tubes (magnetrons, TWTs, or klystrons) generate defective (faulty) current pulses during continuous operation. However, newer tubes may generate defective current pulses at the startup time before they get stabilized. In current research

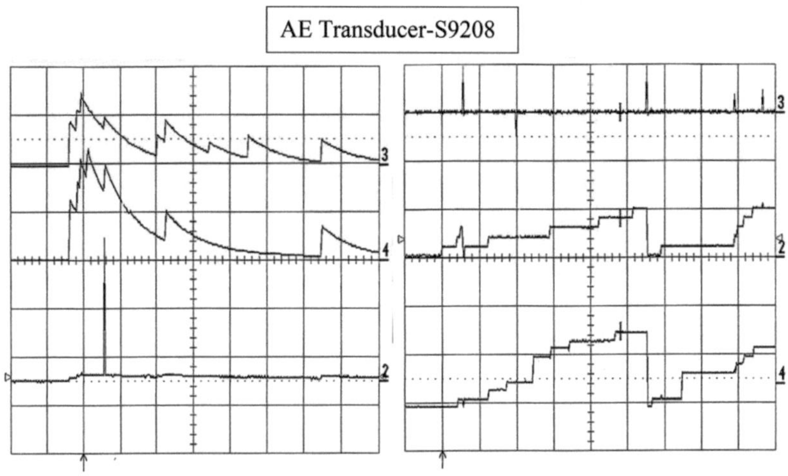

Figure 9.16.
(A) Input to [5C] and output from [5C] and output of [8A] coincide. Ch 2: input to defective current pulse counter. Time scale = 1 s/div. Ch 3: output of defective current pulse counter. Time scale = 1 s/div. Ch 4: output of defective acoustic pulse counter. Time scale = 1 s/div. Ch 3 and 4: vertical sensitivity = 5 V/div. Ch 2: vertical sensitivity = 1 V/div. (B) Outputs from two counters coincide with reset pulses. Ch 3: connected to reset from 1 s. Data sample and reset box [9A]. Ch 2: output of defective acoustic pulse counter box [8A]-counts of bad pulses. Ch 4: output of defective current pulse counter box [5C]-counts of bad pulses. Ch 2, 3, and 4: time scale = 0.2 s/div. Ch 3: reset pulses-separated by 1 s. Ch 3: vertical sensitivity = 1 V/div. Ch 2 and 4: vertical sensitivity = 1 V/div.

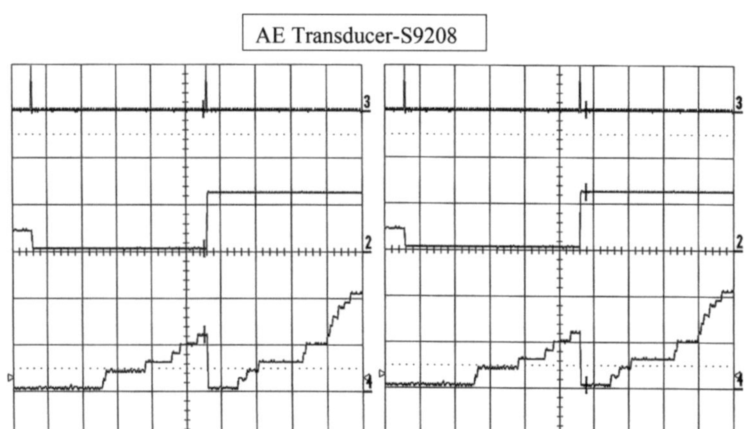

Figure 9.17.
Defective current pulses output transfer. Checking of function of one second data sample and reset box [9A]. (A) Before transfer. Ch 4: signal amplitude at cursor = 609 mV. Ch 2: signal amplitude at cursor = 31 mV. Time scale = 0.2 s/div on all channels of both figures. Ch 3: vertical sensitivity = 1 V/div for both figures. Ch 2 and 4: vertical sensitivity = 0.50 V/div for both figures. Ch 3: to reset on box [9A]. Ch 4: current pulses input to box [9A]. Ch 2: current pulses output from [9A]. (B) After transfer. Ch 4: amplitude at cursor = 31 mV. Ch 2: amplitude at cursor = 625 mV. Ch 3: to reset on box [9A]. Ch 4: current pulses input to box [9A]. Ch 2: current pulses output from [9A].

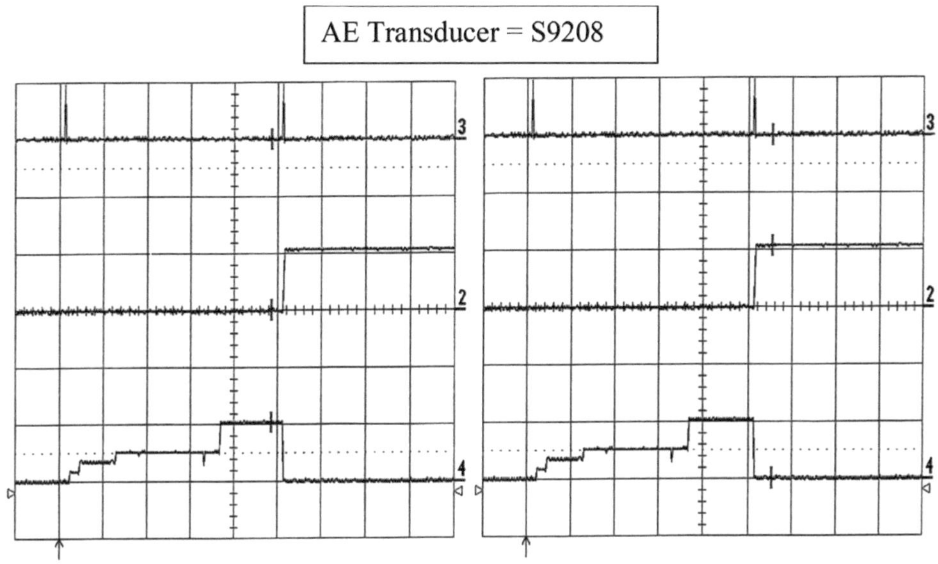

Figure 9.18.
Defective acoustic pulses output transfer. Checking of function of one second data sample and reset box [9A]. (A) Before transfer. Ch 4: signal amplitude at cursor = 516 mV. Ch 2: signal amplitude at cursor = 0 mV. Time scale = 0.2 s/div on all channels of both figures. Ch 3: vertical sensitivity = 1 V/div for both figures. Ch 2 and 4: vertical sensitivity = 0.50 V/div for both figures. Ch 3: to reset on box [9A]. Ch 4: acoustic pulses input to box [9A]. Ch 2: acoustic pulses output from [9A]. (B) After transfer. Ch 4: amplitude at cursor = 0 mV. Ch 2: amplitude at cursor = 531 mV. Ch 3: to reset on box [9A]. Ch 4: acoustic pulses input to box [9A]. Ch 2: acoustic pulses output from [9A].

work, the procedure of restarting the magnetron was used to generate defective current pulses. The faulty current pulses and the corresponding faulty AE pulses were displayed on the oscilloscope along with other related outputs from the other electronic circuits. The results of these tests were recorded as screen captures of oscilloscope displays showing the various signals. The selected displays are shown in **Figures 9.15–9.18** to highlight the performance of the various electronic circuits, and finally, of the whole system. One can find the details of the functioning of electronic circuits in the research report by Joshi et al. [2].

9.5. Experimental results

9.5.1 Display No. 1

The output of the acoustic energy integrator, after passing through the 10-kHz low-pass filter (Block [5A] of **Figure 9.7**) was connected to Channel 3 and to the input of the time delay amplifier and sample-and-hold circuit (Block [6A] of **Figure 9.7**). The output of this second circuit was connected to Channel 4. Defective current pulses were generated by the pulsed magnetron unit, using the previously described restarting method. Channel 3 recorded one defective acoustic energy pulse located at the second vertical division, in **Figure 9.15A**, with the corresponding TTL output on Channel 4.

9.5.2 Display No. 2

The screen display in **Figure 9.15B** shows simultaneously the defective current pulse on Channel 2, defective acoustic energy pulse on Channel 3, and TTL output pulse from the output of the second 10-kHz low-pass filter (Block [7A] of **Figure 9.7**) on Channel 4. All three pulses arrived at the same time, showing the satisfactory performance of these electronic circuits.

9.5.3 Display No. 3

In **Figure 9.16A**, Channels 2 and 3 of the oscilloscope were connected to the input and output of the defective current pulse counter (shown as Block [5C] of **Figure 9.7**). Channel 4 was connected to the defective acoustic pulse counter (Block [8A] in **Figure 9.7**). All three outputs coincide in time, indicating good performance of the electronic circuits involved.

9.5.4 Display No. 4

Next, the functioning of the reset pulses was tested. These pulses occurred at 1-s intervals. The output of the defective acoustic pulse counter (Block [8A] of **Figure 9.7**) was connected to Channel 2, as shown in **Figure 9.16B**. The output of the defective current pulse counter (Block [5C] of **Figure 9.7**) was connected to Channel 4. The resets for both counters were the same. Channel 3 was connected to the reset from the "One Second Data Sample and Reset" circuit (Block [9A] of **Figure 9.7**). Channel 3 shows two reset pulses separated by 1 s. Both counters showed their output counts resetting in step with the reset pulses, as expected. The oscilloscope's voltage measurement cursor is visible as a short vertical line next to the sixth vertical division on all three channels.

9.5.5 Display No. 5

Here, the test of the functioning of the circuit "One Second Data Sample and Reset" (Block [9A] of **Figure 9.7**) was carried out. The maximum output of the circuits for both the defective current pulse counter and defective acoustic pulse counter was 10 V, which was the limit to the DC power input voltage that the analog module [10] could receive. The reset was adjusted to repeat after 99 pulses (99 p × 0.1 V/p = 9.99 V). First, the test was performed for defective current pulses, with the output of the defective current pulse counter (Block [5C] of **Figure 9.7**) connected to Channel 4 and to the input of the "One Second Data Sample and Reset" circuit. The counts are in the form of increasing steps, as shown in **Figure 9.17A**. The output of the "One Second Data Sample and Reset" circuit was connected to Channel 2. The reset pulses are displayed on Channel 3. The voltage measurement cursor appears as a short vertical line on these three channels between the fifth and sixth vertical divisions.

9.5.6 Display No. 6

In this test, the cursor was moved to the right to compare the voltages after the data transfer and reset with voltages acquired previously. **Figure 9.17B** shows that the signal value on Channel 4 was changed from 609 to 31 mV and that a simultaneous change occurred on Channel 2 from 31 to 625 mV. As before, Channel 3 shows two of the reset pulses.

9.5.7 Display No. 7

The data display in **Figure 9.18A** is similar to that of **Figure 9.17A**, but was obtained for the defective acoustic pulse count and the corresponding DC voltage output before transfer and reset.

9.5.8 Display No. 8

In **Figure 9.18B**, the cursor was moved to the right to measure the effects of the acoustic data transfer and reset. Comparing **Figure 9.18A** and **B** shows that the signal value on Channel 4 was changed from 516 to 0 mV and that a simultaneous change occurred on Channel 2, from 0 to 531 mV. As before, Channel 3 shows two of the reset pulses.

After checking the functioning of the "One Second Data Sample and Reset" circuit for the defective current pulses and the defective acoustic pulses separately, the circuit was tested using simultaneous current and acoustic inputs and outputs with simultaneous data transfers and resets. Using a special feature of the LeCroy oscilloscope, two cursors were set on each of the four channels to see the sequence of the data transfer from the "One Second Data Sample and Reset" circuit, shown as Block [9A] in **Figure 9.7**, to the next circuit, the Opto 22 Analog Input Module, shown as Block [10] in **Figure 9.7**. The results of this set of tests are discussed in the next chapter.

9.6. Conclusions

The step of interfacing the outputs of the current sensor and acoustic emission sensor with the host computer must be done using the Opto 22 intelligent digital and analog I/O (input and output) hardware units. The Opto 22 analog input module has a range of 0–10 V for its two inputs. The electronic circuits discussed in Section 3 of this chapter were designed and built during this research project to convert the outputs of the current and acoustic emission sensors into DC voltages compatible with the inputs of the Opto analog input module. We checked these circuits' functional outputs at each successive stage. Finally, we connected and operated the whole system. We then operated the pulsed magnetron intermittently to generate defective current pulses from it, along with defective acoustic energy pulses. The electronically processed outputs from these two sensors were finally received in the OptoScan window of the Opto 22 software. Thus, the total system, composed of the magnetron; the AE equipment; the electronic circuitry of Section 3; the Opto 22 SNAP B3000 brain; and the host computer with the OptoScan software, was assembled and its functioning was tested successfully.

The OptoScan software has only limited capability and cannot be used for constructing virtual sensors. The next step was to develop a system for field applications. For this purpose, ICAS software, currently used to monitor mechanical devices such as pumps and turbines, was more versatile than OptoScan software. The Sensor Editor of the ICAS software translates raw data from an external source into an ICAS sensor value. This process allows ICAS to merge various input types into a single homogeneous "virtual" type. When ICAS receives inputs and converts them to virtual data, all sensors values are alike, regardless of where they originate. Since all sensor values have common units, once inside ICAS, operations between dissimilar devices are possible. The sensor definition elements describe the properties associated with a sensor, all of which can be edited using the Sensor Editor. These properties include data acquisition sources, display units, background scan rates, sensor value calibrations, and many others. The integration of ICAS software with the current system is described in the next chapter.

References

[1] Optomux Protocol User's Guide (Form 203-001026-October, 2000). 43044 Business Park Drive, Temecula, CA 92590-3614: Opto 22 Corporation. Internet website: www.opto22.com

[2] Joshi NR, Ramirez AD, Brock DW, Russell SD. Monitoring of high-power microwave tube systems using the integrated condition assessment system (ICAS). In: Technical Report. San Diego, San Diego, CA: SPAWAR Systems Center; July 2002

Chapter 10

Interfacing and Integrating Acoustic Emission Data (Part II)

10.1. Introduction

The focus in this chapter is on the procedure and details for coupling the acoustic emission-based nondestructive evaluation (NDE) system with the versatile Integrated Condition Assessment System (ICAS) software. This software is used for comprehensive monitoring of many types of equipment including radar systems. ICAS is a comprehensive, maintenance-enabling tool. It provides data acquisition and display, equipment analysis, diagnostic recommendations, and decision support information to plant operators and maintenance personnel. Three primary process applications (software programs) work together to operate ICAS: the main shell, the router, and the analysis engine. Many secondary applications (referred to as applets) enhance ICAS by presenting information, making calculations, performing analyses, and offering advice via Hybrid Decision Support (HDS). ICAS applications receive inputs externally from sensors mounted at various locations on the monitored equipment. These inputs, in conjunction with the ICAS software, work in a hierarchical relationship to perform as designed and to offer a complete comprehensive condition-based monitoring tool.

10.2. Acoustic emission to detect faulty RF pulses

The project proved experimentally that the advanced acoustic emission technique can be used as the nondestructive testing method for in-situ performance monitoring of high power microwave radar tubes like pulsed magnetrons, pulsed TWTs, and pulsed klystrons. It was shown experimentally that the changes in the amplitude and frequency contents of current pulses of the operating microwave tubes were accompanied by the changes in the energy of acoustic emission pulses. Not all RF pulses emitted by a microwave tube such as a magnetron are good. For example, out of 60 pps, 15–20 may be bad. The acoustic emission decreased considerably during the emission of a bad RF pulse from a microwave tube. In this project, electronic circuitry was developed to count the number of faulty pps in the beam current and the number per second of the corresponding acoustic emission from faulty pulses. It was expected that the number of faulty acoustic emission pulses detected in a second would be less than or equal to that of faulty current pulses since the acoustic emission component of the whole experimental setup is less prone to electrical noise compared with the detection circuitry of current faulty pulses.

The next step was to interface the outputs of the current sensor and acoustic emission sensor with ICAS (Integrated Condition Assessment System) software currently used for comprehensive monitoring of many electrical and mechanical systems simultaneously. The interfacing has to be done using the hardware of OPTO22 Intelligent digital and analog input and output (I/O) units. The OPTO22 analog input module takes 0–10 V for its two inputs. The challenge was to develop the electronic circuitry to convert the outputs of the current and acoustic emission sensors into DC voltages compatible with the inputs of the OPTO analog input module. As described in the earlier chapter, this task was successfully completed, and circuits were checked for their outputs at successive stages. The block diagram of these circuits is shown in **Figures 9.2** and **9.7** of Chapter 9. In both figures, the last box (9A) produces counts of both faulty current pulses and faulty acoustic emission pulses. The two outputs of the box (9A) serve as inputs to the analog module (box10) of OPTO22 (OPTOmux) system.

After the whole system was connected together, the system was operated intermittently in order to generate defective current pulses from the pulsed magnetron which in turn generated defective acoustic emission energy pulses. The electronically converted outputs from the two sensors were finally received in the OPTOSCAN software and were displayed on the computer screen. **Figure 9.14** of Chapter 9 shows the OPTOSCAN window. The window shows the baud rate (19200), the Com Port 3, and address 02. On the right hand side, the first two channels (00 and 01) are showing voltage values. The OPTOSCAN software has only limited capability and cannot be used for constructing virtual sensors. The ICAS software has that capability. It was therefore decided to explore the full potential of the ICAS software to increase the efficiency in the field applications of the acoustic emission technique as a built-in test for high power microwave tubes.

10.3. Experimental setup

The complete experimental setup (**Figure 10.1**) assembled could be divided into four components as follows:

The first component of the experimental setup included the magnetron with a current sensor coupled to its cathode cable and an acoustic emission sensor coupled to its anode cylinder, along with the acoustic emission preamplifier and postamplifier and a modulator to provide high-voltage pulses to power the magnetron. The LeCroy Model 9354AM digital oscilloscope was used to observe the various signals. The output (−10 V pulses) of a pulse generator (Interstate Electronic Corporation Model P25) was connected to the trigger input of the magnetron modulator. A 60-Hz synchronizer circuit was connected to pulse generator, which enabled the triggering to be controlled by the on/off switch on the box.

The second component of the setup comprised the different electronic circuit boxes which would convert raw data.

The OPTO22 (Optomux) unit with its power supply, the analog input module, and the brain (A/D conversion unit) constituted the third component of the setup.

The fourth component is the host computer (PC) with ICAS software and OptoScan software.

These four components will now be discussed in more detail.

A. The first component is the high-power magnetron microwave radar tube, with current and acoustic emission sensors with their associated circuitry putting out raw data about both good and bad RF pulses.

Proper operation of the magnetron depended upon the amplitude of the voltage pulse applied and the proper loading of the magnetron by the RF system. Although the magnetron's oscillation frequency was fixed at the design value, oscillation could occur at other than the correct frequency, resulting in overheating and lower output. This is known as double moding and is usually due to a change in loading, or in magnetic field or in the applied voltage. It may be overcome by changing either

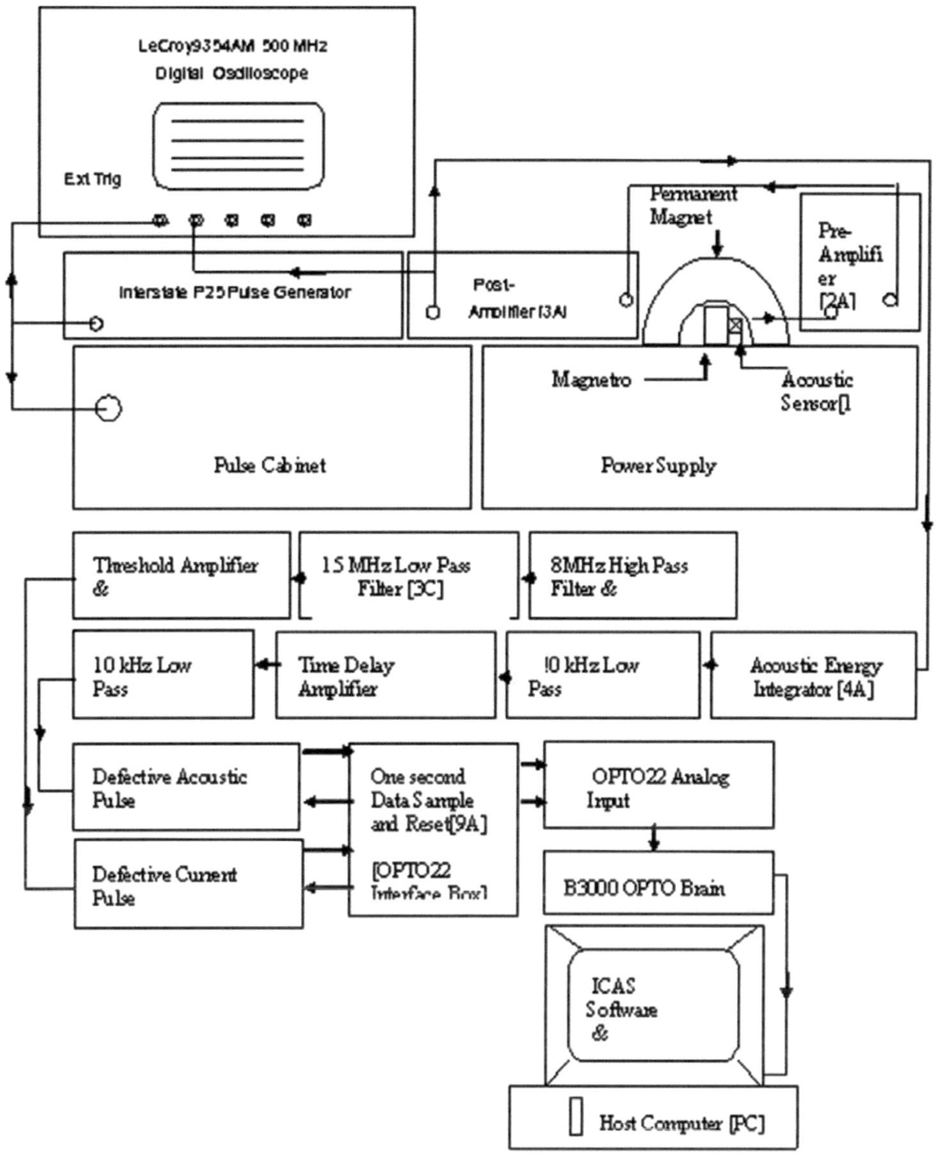

Figure 10.1.
Complete system for counting of current faulty pulses and acoustic emission faulty pulses.

the loading or the applied cathode voltage to obtain a more stable operating point. The outside of the magnetron is a grounded shell to which the plate of the magnetron is connected. The cathode of the tube is heated indirectly and connected to one side of the filament. The filament is fed by a transformer, which is highly insulated to withstand the peak modulator pulse voltage applied to the cathode.

The objective of the test was to detect and count the faulty pulses generated by the magnetron. Some of the causes of faulty functioning of the magnetron are discussed above. Faulty pulses could be generated during the startup of the magnetron before a stable operating point is reached. They could also be generated by reflection of RF energy in the attached waveguide. Additionally, they could be generated by turning the trigger pulses on and off (negative 10 V pulse generated by the Interstate Electronics P25 pulse generator) to the high-voltage modulator that powered the magnetron.

The current and acoustic emission sensors.

The current sensor used in this setup was a transformer, the secondary of which puts out 1 volt per ampere of current flowing to the cathode during pulsing of the magnetron. As discussed in Section 3 of Chapter 9, if the magnetron is putting out a bad pulse, the output voltage pulse from current sensor will have considerable high frequency content. If the magnetron is putting out a good RF pulse, the current pulse will not have the high frequency content. This setup provides a way to identify and count faulty pulses.

The acoustic emission transducer used in the setup was a Model **S9208** (from Physical Acoustics). Its output fed into a preamplifier (2/4/6 Physical Acoustics) and postamplifier (AE1A) with gain settings set at 20 and 21 dB, respectively. As described in Section 2 of Chapter 5, the transducer was coupled to the magnetron with a custom adapter and a coupling compound.

B. The second component—electronic circuits for conversion of raw data from sensors

As discussed in Section 3 of Chapter 9, the output of the current sensor passes through multiple electronic circuits and emerges as voltages representing the defective current pulse count during each second. Similarly, the output of the acoustic emission sensor also passes through multiple electronic circuits and emerges as voltages representing the defective acoustic pulse counts. The outputs of the electronic circuit (Block (9A) of **Figure 9.7)** called "One second Data Sample and Reset circuit" or the OPTO22 Interface circuit were fed to the analog input module.

C. The third component—OPTO22 Optomux I/O unit.

The two outputs of the box (9A) of **Figure 9.7** were fed to the analog input module of the Optomux unit, Block (10). The module has two channels that accept analog voltages ranging from −10 to +10 V DC. The range 0–10 V was used in this setup. The Optomux unit sampled these inputs once per second. In the experimental setup during 1 s, 60 RF pulses are generated, with a mixture of good and faulty pulses. Only the faulty pulses were counted through the interfacing electronic circuitry. The maximum number of faulty pulses per second was 60. The maximum possible input voltage to the analog module was 10 V. This indicates that the voltage increment per faulty pulse should not exceed (1/60) × 10 V or 167 mV. Therefore, it is necessary to check the outputs of the electronic counters, shown as Blocks (5C) and (8A) in **Figure 9.7**, that feed into Block (9A), the OPTO22 Interface box, to determine their output voltage increment per bad pulse.

This was done first for the section of circuit (5C) that counted the faulty current pulses. A square wave from the Interstate Electronics Corporation P25 Pulse Generator was fed into the input of the circuit (5C) and also to Channel 3 of the LeCroy oscilloscope. Channel 1 was triggered by an external pulse source (output to counter reset) coming from circuit (9A). Channel 2 was connected to the current faulty pulse counter output of circuit (5A). The number of pulses per second at the input was increased in an approximately logarithmic manner, and the output voltage was recorded. The experiment was repeated for the section for counting faulty acoustic pulses, with its output connected to Channel 4 of the oscilloscope. The results are given in **Table 1.**

The average voltage increment per pulse was not equal for the current and acoustic counter outputs. Therefore, the gains for these two circuit components were adjusted to give the same value of 0.24 V/p. With this new gain, a count of 42 would give an output of 10 V. Here, the assumption was made that not all 60 pulses during 1 s were going to be faulty. With a gain of 0.24 volt per bad pulse, the output

Calibration of Current Faulty Pulse Counter [box 5C]		Calibration of Acoustic Faulty Pulse Counter [box 8A]	
No. of Pulses/sec Input	Output Voltage in Volts	No. of Pulses/sec Input	Output Voltage in Volts
2	0.47	2	0.62
3	0.78	3	0.94
4	0.94	4	1.25
5	1.1	5	1.56
10	2.2	10	2.97
15	3.3	15	4.4
20	4.2	20	5.47
25	5.2	25	7.20
30	6.3	30	8.4
40	7.5	40	10.5
50	9.2	50	12.7
Total pulses=204	41.19 volts	Total pulses =204	56.01
Ave. voltage per pulse = 0.2019 volts		Ave. voltage per pulse = 0.2746 volts	

Table 1.
After calibration, the gains for both circuit boxes (5C) and (8A) were adjusted to get the same (0.24 V/pulse) output voltage per pulse.

will be 14.4 V if all 60 pulses in a second turned out to be bad. The final operational amplifiers in the counter circuits could actually go to a maximum of 13 V. This was another reason for choosing this value for the gain.

The Optomux I/O unit communicates with the host computer over the RS-485 serial communication link. The Optomux I/O unit converts the analog data into digital data by converting its analog voltage inputs into digital data and then sending the data to the AC37 adapter card installed inside the host computer, over an RS-485 two- or four-wire communication link. Before putting the AC37 card into operation, its jumpers needed to be properly set. It had seven address jumpers (A3 through A9). These jumpers were used to select the base address of the adapter card. In this test, they were set to A4, A5, and A7, giving a base address as HEX 348. The AC37 card allowed for the selection of the interrupt request line with a group of jumpers labeled "IRQ GROUP." Only one jumper was set, on IRQ4. The communication port chosen was COM3. In "GROUP C," consisting of nine jumpers, they were set on the COM ports from COM0 to COM8. In the left-hand bottom corner of the AC37 adapter card, a jumper was set to select OPT0, leaving OPT1 unselected. When a jumper was installed, it sets its jumper value to zero. With no jumper installed, the value would be one. The details of the Optomux protocol used by the OPTO22 I/O unit and of the AC37 card can be obtained from [1].

D. The fourth component—the ICAS software and analysis engine

The ICAS software took data from the Optomux unit and displayed it on the computer screen. However, certain steps were needed to inform ICAS about the sensors and their outputs. These steps were shown in different windows of ICAS, which were used sequentially. Each window needed to be opened in the proper sequence, the required values entered, then executed, and saved. These steps are described in detail in the ICAS manuals supplied by Idax Inc. [2, 3]. The values of parameters seen in the different ICAS windows in the manuals are not necessarily those used in this experimental work.

10.4. Experimental data collection procedure

Following the setup procedure of the previous section, all four components were connected together at this stage of the experiment, so that data could flow from the sensors to the ICAS software installed on the host computer. The oscilloscope used in this procedure was a LeCroy Model 9354AM 500-MHz digital unit with four input channels, each having a sampling rate of 500 MS/s, and 8-bit A/D conversion. Channel 1 is connected to the trigger source (output to counter reset) to display the trigger pulses. Channel 2 shows the output of the counter for faulty current pulses (0.24 V/p). Channel 4 shows the output of the counter for faulty acoustic pulses (0.24 V/p). The time scale of the horizontal time axis was set at 10 s/div. To display data on this long-time scale, the oscilloscope was set into the scrolling mode. When

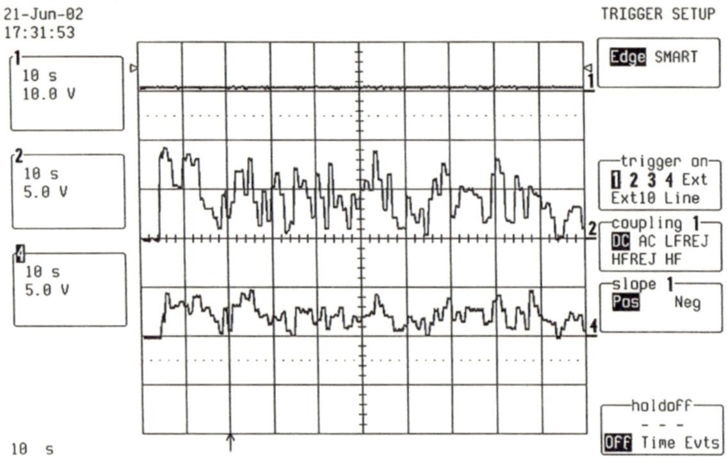

(a)

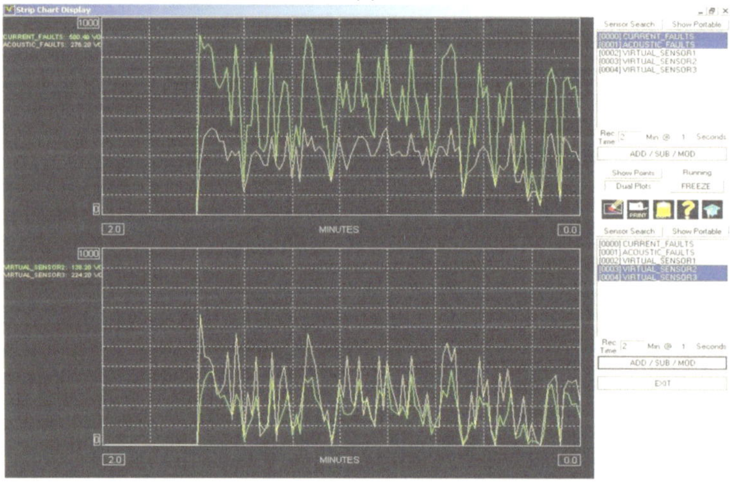

(b)

Figure 10.2.
(a) Faulty pules count per second. Ch 1: trigger pulses-one pulse per second coming from output to counter reset. Ch 2: current faulty pulses per second. Ch 4: acoustic emission faulty pulses per second. Gain on the two counters = 0.24 volts per pulse. On the ninth vertical division on Ch 2, the height of 5 V = 21 bad current pulses per second out of 60 pulses/s. On the ninth vertical division on Ch 4, the height of 3.6 V = 15 bad acoustic pulses per second out of 60 pulses/s. Right edge = zeroth division. (b) Upper display of real sensors-green trace-current faulty pulses/s (X). Yellow trace-acoustic emission faulty pulses/s (Y). Lower display of virtual sensors-yellow trace is (X-Y) and green trace is (X × Y)/1000.

the traces of all three channels reached the left edge of the oscilloscope screen, the scrolling was stopped, and the screen data was saved on a disk in the oscilloscope. The faulty pulses were generated repeatedly by using one of the three methods described above. **Figure 10.2.** shows one such oscilloscope screen display stored on the computer disc.

On the oscilloscope display, the time is zero at the left edge and 100 s at the right edge. Each oscilloscope screen display that was stored on the disk was accompanied by a copy, stored in the computer, of the display of the simultaneous output of the ICAS software. An example is shown in **Figure 10.2.A** and **B**, where **Figure 10.2.** from the oscilloscope is accompanied by **Figure 10.2.B** from ICAS. **Figure 10.2.A** and **B** are both dual chart displays. The upper chart in **Figure 10.2.B** shows the counts of the faulty current pulses during each second, represented by the green color trace, and the counts during each second of the faulty acoustic pulses, represented by the yellow color trace.

The horizontal time scale was then set to 12 s/div, giving 2 min for the whole span of 10 divisions. The ICAS software does not allow a total time for the X-axis of less than 1 min. This time scale was chosen to make the total times on the horizontal axes of both the ICAS and oscilloscope screen displays match each at least approximately. On the vertical axis of **Figure 10.2.B**, each division is 100 V.

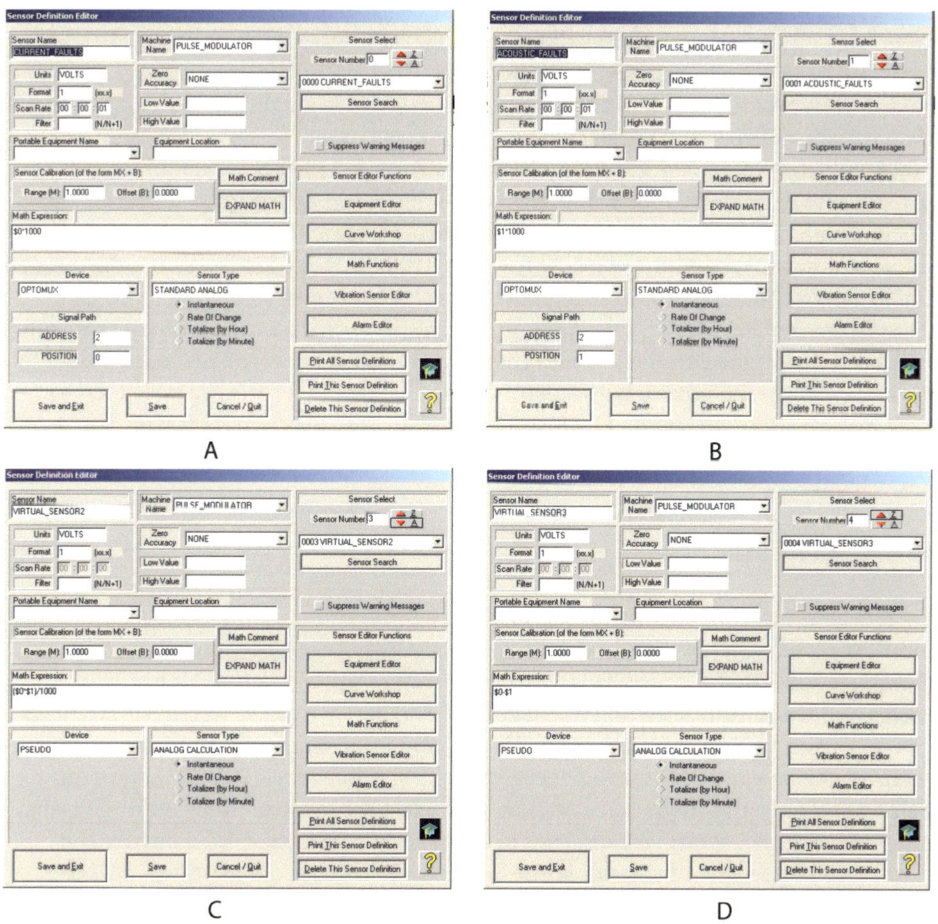

Figure 10.3.
(A) Defining the first real sensor in ICAS window (current faulty pulses/s). (B) Defining the second real sensor in ICAS window (acoustic emission faulty pulses/s). (C) Defining the virtual sensor ((\$0 × \$1)/100) in ICAS sensor definition window. (D) Defining the virtual sensor (\$0-\$1) in ICAS sensor definition.

Figure 10.3A–D shows computer screen displays from the ICAS software that were used to define real and virtual sensors. **Figure 10.3A** and **B** shows the settings to define real sensors for faulty current and acoustic pulses, while **Figure 10.3C** and **D** shows settings to define virtual sensors. In the sensor definition editor window of **Figure 10.3A**, in the math expression box, the output value ($0) of the real sensor0 (current fault count, with the data normalized by dividing by 10, to have a maximum value of 1.0) is multiplied by 1000. Similarly, in **Figure 10.3B**, the output value ($1) of the real sensor1 (acoustic fault count, similarly normalized) is multiplied by 1000 so that in both cases one gets 1000 V for the maximum 10 V count outputs. The values shown in the top left corner of the upper display of **Figure 10.2B** shows the voltage values on both traces at the right edge since the traces were scrolling from right to left.

One of the advantages of the ICAS software is that one can create virtual sensors by combining mathematically the values from real sensors, in this case X and Y. The lower display of **Figure 10.2B** shows plots of values from two virtual sensors. The first virtual sensor plotted is (X-Y), as a yellow trace showing the difference value of current fault counts minus acoustic fault counts. Thus, the difference (500.40–276.20) between the two values at the right edge is equal to 224.20. In this figure, the values appearing in the top left corners of the display again give the values at the right edges of the plots. On both the upper and lower displays of **Figure 10.2B**, the horizontal and vertical scales are the same. The green trace in this figure shows the second virtual sensor, the product (X*Y) of (current fault counts × acoustic fault counts)/1000. For this virtual sensor, the values shown in the top-left corner of the display give a value at the right edge of (500.40 × 276.20)/1000 = 138.21.

Usually, the values indicated by the yellow trace for the acoustic fault count in the upper display are lower than the values displayed by the green trace for the current fault count. This was expected, as discussed previously. **Figure 10.3C** shows the sensor definition editor window for the virtual sensor of ($0 × $1)/1000. **Figure 10.3D** shows the sensor definition editor window for the virtual sensor ($0-$1). The virtual sensors chosen here are only for the illustration of the versatility of the ICAS software. One could decide on the proper combinations of real sensor data to generate virtual sensors, depending on the requirements for the application. Other information about operating ICAS systems using real and virtual sensors can be obtained from the report by Joshi [4].

10.5. Conclusions

The dual (current and acoustic emission) sensor system for in-situ performance monitoring of high power microwave tubes can be successfully integrated with ICAS software used for performance monitoring of on-board machinery. The Integrated Condition Assessment System (ICAS) software is powerful software and has new features like virtual sensors. Depending upon the application, its various features can be exploited for the decision making procedures. It is necessary to delineate clearly the advantages of each of the two sensors and to prove under what circumstances AE sensor is preferred to electric current sensor. To compare the difference between functioning of the two sensors, additional experiments need to be arranged.

Sensors are the input legs of ICAS. Sensors form the boundary between the real world and the virtual environment that ICAS uses, once it has normalized all sensor information. ICAS monitors and predicts machinery or equipment failure modes by comparing the sensor data to established engineering performance criteria. If a machine's actual performance (as compared to its design performance) violates a specified limit, an alarm is activated. To utilize this ability of ICAS, data generated at specific intervals of time during the in-situ performance monitoring of high power microwave tube AE techniques need to be stored in the ICAS database. The task of storing the data, comparing the data, displaying the trend indicator graphs, and setting alarm activation guidelines is the next step to be undertaken to complete the integration. Since ICAS has ability to take data inputs from multiple sensors, the multi-sensor, multi-channel acoustic emission system in place of the present single sensor system could also be used. ICAS can use calculated or pseudo (virtual) sensors in addition to the physical sensors.

In conclusion, ICAS can be an effective diagnostic tool for high-powered microwave tubes by using data from multiple sensors, comparing data with reference values to aid in decision-making procedures and creating new virtual sensors by combining data from real sensors.

Author details

Narayan R. Joshi[1], Ayax D. Ramirez[2*], Stephen D. Russell[2] and David W. Brock[2]

1 Independent Scholar, USA

2 Space and Naval Warfare Systems Command (SPAWAR), USA

*Address all correspondence to: ayax.ramirez@navy.mil

References

[1] OPTOMUX Protocol User's Guide (Form 203-001026-October, 2000). 43044 Business Park Drive, Temecula, CA 92590-3614: OPTO22 Corporation. Internet web site:www.opto22.com

[2] ICAS-CDS Developer Manual. 5301 Robin Hood Road, Suite 134, Norfolk, Virginia, USA: IDAX Incorporated (www.idax.com)

[3] ICAS-System Administrator User Manual. 5301 Robin Hood Road, Suite 134, Norfolk, Virginia, USA: IDAX Incorporated (www.idax.com)

[4] Joshi NR. ICAS system to monitor performance of microwave radar power tubes with acoustic emission technology (Part IV-Magnetron). Technical Note 1885. San Diego, Advanced Technology Branch, Code 2853: SPAWAR Systems Center; July 2002